General Aviation Law

Other Books in the Practical Flying Series

Handling In-Flight Emergencies by Jerry A. Eichenberger
Cockpit Resource Management: The Private Pilot's Guide by Thomas P. Turner
The Pilot's Guide to Weather Reports, Forecasts, and Flight Planning, 2nd Edition by Terry T. Lankford
Weather Patterns and Phenomena: A Pilot's Guide by Thomas P. Turner
Cross-Country Flying by Jerry A. Eichenberger
Avoiding Mid-Air Collisions by Shari Stamford Krause, Ph.D.
Flying in Adverse Conditions by R. Randall Padfield
Mastering Instrument Flying, 2nd Edition by Henry Sollman with Sherwood Harris
Pilot's Avionics Survival Guide by Edward R. Maher
The Pilot's Air Traffic Control Handbook, 2nd Edition by Paul E. Illman
Advanced Aircraft Systems by David Lombardo
The Pilot's Radio Communications Handbook, 4th Edition by Paul E. Illman
Night Flying by Richard F. Haines and Courtney L. Flatau
Bush Flying by Steven Levi and Jim O'Meara
Understanding Aeronautical Charts, 2nd Edition by Terry T. Lankford
Aviator's Guide to Navigation, 3rd Edition by Donald J. Clausing
Learning to Fly Helicopters by R. Randall Padfield
ABC's of Safe Flying, 3rd Edition by J. R. Williams
Flying VFR in Marginal Weather, 3rd Edition by R. Randall Padfield
The Aviator's Guide to Flight Planning by Donald J. Clausing
Better Takeoffs and Landings by Michael C. Love
Aviator's Guide to GPS, 2nd Edition by Bill Clarke

General Aviation Law

Second Edition

Jerry A. Eichenberger

McGraw-Hill

New York San Francisco Washington, D.C. Auckland Bogotá Caracas Lisbon London Madrid Mexico City Milan Montreal New Delhi San Juan Singapore Sydney Tokyo Toronto

Library of Congress Cataloging-in-Publication Data

Eichenberger, Jerry A.
General aviation law / Jerry A. Eichenberger.—2nd ed.
p. cm.
Includes bibliographical references and index.
ISBN 0-07-015104-0 (hardcover).—ISBN 0-07-015103-2 (pbk.)
1. Aeronautics—Law and legislation—United States—Popular works.
I. Title.
KF2400.Z9E36 1997
343.7309'7—dc20
[347.30397] 96-34969
CIP

McGraw-Hill
A Division of The ***McGraw-Hill*** *Companies*

8 9 10 11 12 13 14 15 DOC/DOC 0 9 8 7 6 5 4 3 2

ISBN 0-07-015103-2 (HC)
ISBN 0-07-015104-0 (PBK)

The sponsoring editor for this book was Shelley Chevalier, the editing supervisor was Fred Bernardi, and the production supervisor was Suzanne Rapcavage. It was set in Palatino by North Market Street Graphics.

Printed and bound by R. R. Donnelley & Sons Company.

Disclaimer: This book is intended, and should be used as, a source of general information only. The contents are not to be relied upon or construed as legal advice. The reader is urged to seek the advice of an attorney-at-law to render advice on any particular concern or transaction. In addition, the law varies from state to state within the United States, reinforcing the need for the services of an attorney in any specific situation.

This book is dedicated to my loving wife Candy
and daughter Sara, with whom I have been privileged to share
not only my life, but also an untold number of hours in the sky,
and without whose heartfelt encouragement and understanding
for the many nights spent in my office writing these pages,
this work would have in no way been possible.

Contents

Preface xi

1 The American Legal System 1

Heritage of the English Common Law 2
State Law Versus Federal Law Relating to Aviation 2
The Functions of Judge and Jury 3
Some Common Myths About Justice 4

2 The Federal Aviation Administration 7

Brief History of the FAA 7
Functions of the FAA Today 8
  Rule Making 8
  Policy Making 10
  Enforcement 10
  Relationship to the NTSB 11
Airman Certificates and Ratings 11
  Student Pilot 12
  Recreational Pilot 13
  Private Pilot 15
  Commercial Pilot 17
  Airline Transport Pilot 19
  Certificated Flight Instructor 21
Currency Requirements and Flight Reviews 24
Maintenance Certificates 26
  Airframe Mechanic 27
  Power Plant Mechanic 27
  Inspection Authorization 28

3 Aircraft Ownership 31

Acquisition of an Aircraft 31
- Prepurchase Inspection 32
- Contract of Sale 34
- Title Search 35
- Aircraft Title Insurance 37
- Bill of Sale 38
- Security Agreements 41
- Application for Certificate of Registration 43

Co-ownership of an Aircraft 43
- Informal Co-ownership Among Individuals 47
- Flying Clubs 51

Owner-Performed Maintenance 54

4 Homebuilt Aircraft 59

Building a Homebuilt—Plan First 61
The 51 Percent Rule 62
Certification Steps 65
Maintaining Homebuilts 66
Operating and Insuring Your Homebuilt 67
Restoring the Older Airplane 72
Selling the Homebuilt 74

5 Aviation Insurance 77

Use Restrictions 77
- Pleasure and Business Coverage 78
- Industrial Aid Coverage 79
- Commercial Coverage 79

Pilot Warranties 80
Hull Coverages 83
Liability Coverages 84
- Per Person Limitations 85
- Combined Single Limit 86
- Renter Pilot Coverage 87

Duty to Defend 89
Subrogation 90
Illegal Use of the Aircraft 91

6 FAA Enforcement Procedures 95

Letter of Investigation 96
- Response to Letter of Investigation 96

Notice of Proposed Certificate Action 98
- Options in Response to Notice of Proposed Certificate Action 100

The Stale Complaint Rule 104
Notice of Certificate Action 105
Emergency Enforcement Actions 107
The Aviation Safety Reporting System 108
Appeal to the National Transportation Safety Board 115
Further Appeals 120
Civil Penalties 121
Administrative Action 123

7 Principles of Negligence Liability 127

Duty to Others 128
Breach of Duty 130
Proximate Cause 133
Damages 134
- Special Damages 134
- General Damages 135
- Preponderance of Evidence 138

Defenses to Negligence 139
- Assumption of the Risk 140
- Contributory and Comparable Negligence 140
- Last Clear Chance 142

8 Particular Applications of Negligence Law 145

Those Who Rent Aircraft to Others 145
Flight Instructors 151
Fuel Suppliers 157
Maintenance Operators 159
Fixed-Base Operators 164
Pilots 173
Flying Clubs and Other Co-ownership Agreements 174

9 Product Liability 177

Negligent Design and Manufacture 179
Warranties and Representations 181
Strict Liability 186
The Consumer Expectation Test 188
The Risk-Benefit Comparison 189
Failure to Warn 193
General Aviation Revitalization Act of 1994 195
Suppliers' Liability 198
Crashworthiness 202

10 Medical Certificate Appeals and Special Issuance 207

Check Yourself First 208
A Request by the FAA for Further Records 209
Waivers and Special Issuance Certificates 210
Appeal of Denial of a Medical Certificate 213

11 Conclusion 217

Index 221

Preface

IN THE YEARS SINCE I BEGAN PRACTICING LAW, THE UNITED STATES aviation industry has been at the center of many of the rapid and significant developments in the American legal system.

I occasionally have the privilege to speak to groups of pilots, fixed-base operators, mechanics, and FAA personnel. Everywhere I present these talks, some of the topics covered in this book draw a good deal of interest. Everyone seems, at the same time, to be mystified, frustrated by, and angry at the law as each person perceives it, as well as at the lawyers whom all people want to hold responsible for everything that is seen as wrong with the system.

Without doubt there are injustices and errors in the workings of the law. So are there in all the other systems and mechanisms that civilization has developed. But the best way to be equipped to deal with the law is through a working knowledge of the parts that affect each person directly, whether it be in his vocation or avocation. That is what this book is about.

There is certainly no intent in these pages to make lawyers of readers. Many would lay the book down immediately if that were the goal, not wanting to join such ranks even if they could. The point of this work is to pass along some working information to those who hold the same affection for flying as do I.

Physicians often have it easy in one respect. Patients usually know when they are sick and go to see the doctor before it is too late. Lawyers often do not share that luxury. Our clients generally do not recognize the onset of a legal problem before serious compromises in their position have occurred, some nigh irreparable.

Inevitably, these folks eventually do find their way to an attorney, who is naturally expected to turn back the clock, ignore all the unfavorable facts, and save their skins. If she does not, she is considered to be incompetent, while at the same time the system is seen as so perverted that no honest person can get a fair result.

Rather than fit into that class of individuals, it is hoped that the reader will know what basic problems to avoid in the first place, and when to seek the counsel of an aviation lawyer—before it is too late.

People pay tens of thousands of dollars for an airplane, yet some rebel at the thought of spending less than an additional $100 for a title search to get some reasonable degree of comfort that they will end up owning the dream machine for which they have just paid so much.

Others have no real concept of what insurance coverages they do and do not have. So long as you have a million or so worth of total liability coverage, good enough, right? We shall see.

Since the late 1980s, many aviation pundits have blamed product-liability litigation for the precipitous decline in the American general aviation industry. There is no doubt that lawsuits against manufacturers have had some effect upon our industry, but we'll take a deeper look at the subject, including analyzing just how much effect the litigation really had, and what has been done recently to lessen the courtroom burden on manufacturers of general aviation products.

What do you do when confronted with a letter from the FAA informing you that an investigation is in progress, concerning a flight you recently made, to see if you violated any Federal Air Regulation (FAR)? The letter invites, even encourages you, the pilot, to respond within a relatively short time so that the person making the investigation will have the benefit of your side of the story. Of course, you do not want to ignore the writer. What next?

Flying is a wonderful way to earn a living, and one of the most enjoyable forms of transportation that there is. It is a hobby that gives untold thousands of hours of relaxation and sport to many aviators each year. The regulatory burden need not be a backbreaking toll. The fear of a minor accident does not have to haunt a pilot's every flight, if he is properly insured. Likewise, pilots live through scrapes with the FAA over alleged violations of the FARs in the same way you probably did with your last encounter with the highway patrol.

This book is not intended to drive anyone away from aviation with its stories. It is simply designed to make each reader a little wiser and more sage in the workings of a system that affects each of us and what we all hold so dear.

Jerry A. Eichenberger

1

The American Legal System

AT FIRST SIGHT, THE AMERICAN LEGAL SYSTEM CAN APPEAR TO BE A maze that has no particular path through it. Lawsuits have results that seem to be incredible. Millions of dollars are reported as being awarded on claims that appear to be frivolous when viewed from the outside, without a thorough and working knowledge of the facts or the law of those particular cases. To the uninitiated, the workings of the law, courts, and lawyers can be mysterious. There is no question that our system is built upon rules that are not easily, if ever, thoroughly understood.

In the past, law students learned of the principle known as *stare decisis*, which means the "law of the binding precedent." Our system developed around the concept that judges would find and apply particular law to particular cases. Then, once decided, those cases could be used in analogous future situations as authority for what the law is. Supposedly, lawyers in practice and judges in courtrooms would then apply that law decided in the former case to the determination of the case presently before them.

This concept of legal precedent began hundreds of years ago and served us well for a long time. The problem today is that as fast as the law is changing and evolving to meet the needs of an even faster-changing society, the value of *stare decisis* has diminished because the courts are growing more reluctant to simply declare that what once was a proper outcome of a particular controversy is fit for present application.

In many areas of our legal system—the first two of which come to mind are the criminal and civil rights arenas—the law changes extremely fast. In many cases, it ought to. To hold today to the principles that permitted the lack of civil rights just a few decades ago would not be acceptable to many of us. Yet in the areas of business and accident law, even though the technologies are developing daily, the legal principles are not. There is not that much difference between a broken wagon wheel and a broken landing gear, as far as the necessity arises to apply the law to an accident caused by that failure. Yet so many writers and practitioners within our legal system seem to think that we need to change the principles that guide us to keep pace with the development of scientific and engineering technology. Perhaps some change is in order in these old areas of the law, but I for one feel that change, simply for the sake of change, is a poor practice.

Let us examine from where our legal system came, how it is applied to aviation cases, and some common myths about it.

HERITAGE OF THE ENGLISH COMMON LAW

The United States had its beginning in the original 13 English colonies, which were quite developed in both their economic and judicial systems at the time of the American Revolution, and by the 1770s the *English common law* was the legal system in use. Then, as now, most of the politicians and political writers of the day were either practicing lawyers or people with a legal education.

After the American Revolution, the framers of the new nation saw no reason to deviate from many of the institutions that they had inherited from the mother country. Other than changing the monetary system from the English pound to the Spanish dollar, virtually all other customs, mores, and principles of the new society remained the same as when the colonies were British possessions. Because courts had existed for well over 100 years before the revolution, and the merchants, lawyers, and judges were familiar with the workings of the English system, which appeared to be accomplishing the purpose of settling disputes, there was no reason to change.

Hence, the concept of a common law system, in which judges sitting deciding cases are responsible for a great measure of the legal principles that guide and govern us, was so deeply ingrained in society that it survived any emotional desire to change it simply because it was English, and has persevered throughout the 200 years of our nation's existence.

Naturally, the American common law system has evolved somewhat differently than the pure English workings upon which it is based. But to a surprisingly great degree, American law today, particularly the judge-made law, is not significantly different from that in Great Britain, or any of its former colonies or possessions such as Canada, India, South Africa, or Australia.

STATE LAW VERSUS FEDERAL LAW RELATING TO AVIATION

Aviation is a unique industry and endeavor in more ways than one. First of all, as industries, activities, and economic systems go, it is a newborn. Because it has been fewer than 100 years since the Wright brothers first gave us powered flight, only a short moment of time has passed, as the legal system defines "time" in terms of how long it takes to work changes in the law.

The American legal system is basically composed of two competing subsystems of law. First and foremost, the *law of the various states* in which we live and do business controls the lion's share of our daily activities. It is state law, not federal law, that governs most of the day-to-day business transactions with which we deal; it is state law that prohibits us from taking our neighbor's property or life, that regulates how we drive our automobiles, that deals with what happens when accidents do occur in terms of who is at fault or liable to compensate another.

Yet we have a system of *federal law* that pervades the nation and emanates from our capital. Unlike the English system previously discussed, our American legal system has created a dichotomy between those areas over which the federal government has jurisdiction, which come from the United States Constitution, and the rest of the activities of society, which are governed and controlled by state law.

Because aviation is obviously a highly mobile undertaking, it quickly came to be viewed as an area appropriate for federal intervention and the application of federal law. There have been several attempts on the part of the federal government to regulate aviation, and the *Federal Aviation Act of 1958*, as amended, is the current general statute passed by Congress that in a comprehensive sense, governs and regulates aviation. We will see in Chapter 8 that Congress passed a law in 1994 which gives a substantial amount of relief from product-liability lawsuits to general aviation manufacturers.

However, there is still a large area of the law affecting flying that is purely the law of the various states. For instance, when an accident does occur, it is generally state, not federal law, that decides if someone is to be at fault or is to be held liable to another, the conditions under which that liability occurs, and the amounts and types of compensation to be paid to the aggrieved parties. If a manufacturer is subject to suit, state law still determines the outcome of the case, once a plaintiff gets past the limitations of federal law, and gets into the courthouse.

Therefore, aviation lawyers are constantly dealing with federal law as well as with the law of the state in which they practice and where the particular accident or transaction occurred, or, in many instances, they are dealing with the law of another state where such happened. When another state is referred to as a *foreign jurisdiction*, the use of that phrase does not indicate a foreign country, but rather simply another state.

Because the 50 states are just that—separate states, each with its own laws, legislature, and courts—it is recognized that they are, in fact, foreign one to the other. The law of Ohio in many situations is quite different, for instance, from the law of New York or the law of California. Should a particular accident or business transaction occur in any of those states, a person must be thoroughly familiar with the law of the state that will control the eventual outcome of the controversy.

Quite often, a lawsuit is tried in one state and the law of another state is used to either decide the entire litigation, or certain steps of it. The question of which law to apply to a given situation is called *conflicts of laws* and is an entire subject matter itself, with complex rules devised to decide the issue of which state's law is to govern and resolve the dispute that is before the court.

THE FUNCTIONS OF JUDGE AND JURY

One of the most inviolate principles of the English common law system that we inherited is that of *trial by jury*. The English long ago realized that they wanted their disputes resolved, to the degree possible, by laypersons rather than by judges who would apply legal technicalities to often subvert concepts of common sense. Hence, the jury system is used in all 50 state legal systems and in the federal courts as well.

Although there are a few types of cases that are not subject to jury trial in most states—domestic relations proceedings are an example—in many states virtually all the lawsuits that arise from aviation accidents and business transactions will be subject to trial by jury. Therefore, it is necessary that we gain an initial understanding of what it is the jury does in the courtroom setting, and what functions are performed by the judge.

Basically, the jury is impaneled to hear the evidence that is presented in a lawsuit and determine questions of fact. This means it is the jury who decides what happened, by or to whom, and generally under what circumstances. It is the jury who decides whether the pilot or operator of the aircraft was operating in accordance with good operating practices, or if the pilot was negligent. It is the jury who has the duty to determine if a product is defective when it does not live up to the standards that the normal consumer would expect of it.

On the other side of the coin, *the judge's role is to determine questions of law*. The judge decides whether particular items of evidence are admissible under the convoluted and complex rules that govern introduction of evidence. It is the judge who sustains or overrules objections directed to the admissibility of documents or testimony into evidence at a trial. Likewise, the judge has the job of determining what law governs the particular controversy and instructing the jury on that law, so that when the jury retires to deliberate its verdict, it can apply the law the judge has given to it and, in combination with its own determination of the facts, render a verdict for one side or the other.

Many times the line of demarcation between a question of fact and a question of law becomes quite blurry. However, simply remember that it is the jury who generally decides if someone is telling the truth and, if so, to what degree. It is the role of the judge to control the courtroom—particularly, but not limited to, reigning in the lawyers—and to see that the procedure, as well as the evidence and arguments submitted to the jury, complies with the legal standards in that judge's jurisdiction.

SOME COMMON MYTHS ABOUT JUSTICE

From the time that all of us were youngsters, and in particular beginning with our school years, we have had ingrained in us the principle that the American legal system is a unique engine designed to ferret out the truth and always accomplish justice. While that is without question the aim and goal, all too often it is not the result.

Ours is an *adversary legal system*. That means that it is the job of a lawyer to advocate a client's position within the bounds of the rules of law and the ethical constraints that guide and govern us. The lawyer's job is to find a legally arguable position that supports the result that the client wants. Without question, the client wants to win. The client is not particularly interested, in most cases, in achieving what might be a fair allocation of responsibility, because each of us has his or her own definition of what constitutes a fair result.

Therefore, our system has, from its beginnings, held to the proposition that the lawyer is engaged to do battle for a particular client. The lawyer is to advocate that party's position, introduce the evidence in a manner that furthers that goal, and then argue the interpretation and application of that evidence to accomplish the desired outcome, which is always that the client will prevail in the litigation process.

In fact, attorneys in our system have the absolute duty to be zealous advocates, again within the bounds of the law and ethical obligations. Most certainly, the attorney is not engaged to misrepresent the evidence, but simply to present it in a manner that benefits the client to the greatest possible degree.

At the same time, a competent attorney should take a somewhat distant and objective view in counseling clients during the process of litigation, particularly, but not limited to, the settling of a dispute. A lawyer should always advise clients using professional judgment and experience as to the likelihood of success of the client's position.

Lawyers often advocate positions that do not, at first sight, seem to have a great likelihood of victory. Many of our great turning points in the law have been the result of such cases. It is not the attorney's job to decide the outcome of the case—that is left to the court.

Therefore, attorneys are often likened unto actors. The manner of the presentation of the evidence, the rapport that a particular trial lawyer can generate with the jury and, to some degree, the judge, can often have a tremendous impact on the outcome of a trial. While not reduced to a purely theatrical performance, modern-day litigation does put the lawyer in a position of often being seen by the jury as one of many determining factors in the eventual result of the trial. This might not be particularly fair, but humanity has never developed a totally fair or perfect system.

It has been said by at least one prominent commentator on the common law system that it is fraught with injustice and unfair results; but that we have yet to develop a better or more civilized means for settling disputes. And in the end, the legal system exists to settle the disputes of those who come under its jurisdiction. As with any other argument, there is often no perfect solution. No one wants to be injured in an accident; no one wants to undergo such devastating trauma—or to see that happen to a family member—let alone meet an untimely death.

The legal system can never heal unending misery, it cannot bring back those killed in accidents, it cannot undo maiming. All it can do is provide a measure of compensation, usually measured in money, to the individuals who feel they have been aggrieved at the hands of another.

Therefore, if particular people are so unfortunate as to be injured, let alone have their lives ended in an aviation accident, true fairness can never result, because humanity is incapable of accomplishing it. Those who feel that they have been deceived, cheated, or simply mistreated in a business transaction can almost never recover all their real or imagined losses.

The legal system has to be seen for what it is: an imperfect yet constantly evolving and, hopefully, improving method of resolving the inequities that one segment of our society works upon another. It is the lawyer's job to attempt, through the advocacy of his client's position, to remedy those happenings as best he can. But we must all realize that we will never achieve perfection in ourselves, nor in the systems that we develop.

When a particular person is involved in the legal system, whether it be as *plaintiff* (the one who is bringing the suit) or as *defendant* (the one who is being sued), she must approach her involvement with a realistic eye and pursue realistic goals. To do otherwise, in the arena of civil litigation, will almost guarantee that she will come away frustrated, angry, and irate at lawyers, juries, and the law in general.

In every courtroom across our country, and in every trial, there is a winner and there is a loser. Reasonable people often can, with the assistance of competent counsel, realize the compromising aspects of their relative positions, and eventually settle their own disputes.

But when that is not possible, for a variety of reasons, the system must perform its time-honored function and settle the dispute for them. Simply remember that such a forced resolution seldom will be a pleasant experience. The result similarly will be seldom exactly what either side desired and, for certain, will be a result that at least one side did not desire.

2

The Federal Aviation Administration

AS CURRENTLY SET UP, THE FEDERAL AVIATION ADMINISTRATION IS A branch of the United States Department of Transportation. The federal administrative agency that has governed aviation since the industry's infancy has gone through several changes, several functions, and several stages of interrelationship with other federal agencies. In order to understand where the FAA is now, we would be wise to spend a few minutes reviewing where it began and how it got where it is today.

BRIEF HISTORY OF THE FAA

When aviation began on that cold December day in 1903, the Wright brothers were completely and totally free to design their airplane as they saw fit, operate it as they pleased, and allow any person to fly it who had the intestinal fortitude to clamor into the contraption and attempt to take wing in it.

Not that much changed between 1903 and the 1920s. As we all know, aviation received a tremendous boost both in popularity and in technical achievement from the years of World War I. It was not until quite some time after the great war that aviation came under any particular regulation or regulatory activity in the United States.

In the late 1920s the Department of Commerce began rudimentary regulatory activities over the civil aviation industry and its operations. At that time in history, the department issued the first pilot certificates. Until then, pilots did not have to be licensed in any manner. No particular displays of competency were required, nor did pilots have to prove that they met any standards of physical fitness. Everyone and anyone flew whenever they pleased, wherever they pleased, in whatever they pleased, and in any weather they had the courage to try. Needless to say, civil aviation was not considered to possess a stellar safety record. All it took for someone to learn to fly was to find somebody willing to teach; the teacher might well have had not a great deal more experience than the student. All it took to build an airplane was the financial wherewithal to purchase the materials and the daring to fly the craft. At about 1928, all that began to change.

The first pilot licensing standards were quite sparse. In fact, the Department of Commerce inspectors were not pilots themselves, and most were fearful of even getting into a flying machine. They simply stood on the ground and watched the candidate take off, circle the field, do a spin or two, and land. If the candidate lived through that experience, the new pilot was then certificated. Approximately the same degree of thoroughness was attached to the certification of airframes and engines in those early days.

The Civil Aeronautics Administration came upon the scene in the 1930s. By that time, most flyers had finally bitten the bullet and submitted themselves to the hated license examination, and some of the first written examinations were given. Gradually, the certification system evolved and various levels of pilot certificates, similar to what we know today, were developed. In those days there were certificates such as the limited commercial certificate and the limited flight instructor certificate, which have long since disappeared.

At the conclusion of World War II, the basic scheme of pilot licensing and airframe and engine certification had developed to the point where it was not greatly different from that in use today.

In 1959, the Civil Aeronautics Administration ceased to exist, and in its place Congress authorized the formation of the Federal Aviation Agency, which was a separate agency, responsible to no particular cabinet-level department. It was felt at that time that independence was necessary to depoliticize the body that regulated civil aviation and to cure some abuses that had occurred during the tenure of the old CAA.

However, the agency concept was not long-lived. In the 1970s, with the general reorganization of the cabinet-level offices in the federal government, the Department of Transportation was formed and the FAA changed its name to the Federal Aviation Administration, under the control and purview of the Department of Transportation. There have been some attempts in Congress to revitalize the separate agency concept, but as of this writing, those measures have not met with a great deal of success, and the FAA continues to be the Federal Aviation Administration.

FUNCTIONS OF THE FAA TODAY

The FAA has three primary functions today, and a fourth that is its interrelationship with the National Transportation Safety Board. The three primary functions are:

1. Making of rules to govern and regulate civil aviation
2. Making of long-term policy toward not only the governance of aviation but also the fostering of it
3. Enforcement of the rules that it makes

Rule Making

It comes as no surprise to anyone that the primary function of the FAA is to regulate civil aviation. In so doing, it proposes, promulgates, and enforces certain titles of the Code of Federal Regulations that are referred to in the aviation industry as the *Federal Aviation Regulations*. The correct way to refer to them, however, is particular parts and sections of *Title 14 of the Code of Federal Regulations*. Whatever label you put upon them, the regulations control and dictate virtually all facets of nonmilitary aviation.

A Federal Aviation Regulation is changed or originated when the FAA issues a document known as a *Notice of Proposed Rule Making*. Federal law requires virtually all the administrative agencies and departments of the federal government that adopt regulations, such as the Occupational Safety and Health Administration, Internal Revenue Service, Federal Communications Commission, and of course the FAA, to conduct their rule-making operations in this manner. These regulations have the force of law, but are unique.

Everyone has had high school civics lessons regarding how the government is organized. The function of the *legislative branch*, namely Congress, is to pass laws; the job of the *executive branch* and its many agencies and administrations is to enforce those laws; and the function of the *courts* is to interpret and apply those mandates. The entire body of *administrative law* is a special area unto itself.

Administrative agencies often act as rule makers, rule enforcers, and, to a great degree, arbiters of conflicts that come about as a result of persons operating in spheres of activity controlled by those rules. The FAA is no exception.

The law requires that when the Notice of Proposed Rule Making (NPRM) is issued, the administrative agency must allow a particular period of time for comment upon that rule, unless certain emergency conditions exist. Therefore, when the FAA wants to either originate a new regulation or change an existing one, it issues the NPRM and allows the industry a reasonable period of time in which to comment. Quite often, the comment period appears short, and various industry sources petition the FAA to extend that comment period, which is frequently accomplished.

After the comment period closes, the agency is then supposed to consider the comments of the public and those to be affected by the rules proposed, and deliberate and consider the same in its process of rule making. Quite frequently, a large outpouring of comment does in fact influence the FAA. Several proposed rules over the past few years have been abandoned or significantly modified after the consideration of public comments. When this process has run its course, the agency then issues its *Final Notice of Rule Making*, which sets forth the rule as it will be adopted, and gives an effective date for it.

There is absolutely no doubt that public comment is a most important stage of the rule-making process and is, frankly, the only one in which the average person to be affected by the rule has any real voice. Although trade associations and other relevant groups frequently meet with representatives of the FAA to discuss upcoming rules, average people on the street have an opportunity to make their feelings known through the public comment process. As will be discussed in a short while, the recent adoption of the Notice of Proposed Rule Making concerning the recreational pilot certificate came about in large measure as a result of the overwhelming public comment supporting the creation of this new level of pilot certificate.

Policy Making

A great part of the mandate given the FAA by the Federal Aviation Act of 1958, as amended, is to foster and support the growth and use of civil aviation in the United States. Unquestionably, the FAA's rule-making activities speak as evidence of current administration policy toward civil aviation.

But there are other ways in which the FAA makes policy. Particularly in the last decade or so, the administrator of the FAA has become a more visible public official, more outspoken in approaching things, and more available to the industry and the media to pronounce those feelings that inevitably reflect the policy of the FAA.

Shortly after the midair collision that occurred over Cerritos, California in the 1970s, the administrator quickly let it be known that the enforcement posture of the FAA concerning violations of the regulations by pilots was going to be more strict. Enforcement is addressed in depth in Chapter 6, but the point here is that through public appearances, press releases to the media, and industry forums, the administrator—along with assistants, regional directors, and other administrative staff—clearly sets forth and announces policy emanating from Washington and the various regions of the FAA.

That is as it should be. None of our governmental agencies, except those engaged in highly sensitive matters of national security, should operate in a vacuum. Notices of Proposed Rule Making should, and generally do, become issued as a result of previously or concurrently announced administration policy.

Enforcement

Chapter 6 delves into detail concerning the process that is used by the FAA and the National Transportation Safety Board to enforce the Federal Aviation Regulations against those who act in violation of them. As the American administrative legal system has developed, it has also become a function of the various administrative agencies to generally act as the enforcement arm to deal with violations of the regulations that they have made and adopted.

Enforcement practice has certainly changed over the years, and we are currently embroiled in a period of high enforcement activity. The FAA's *enforcement* of its own regulations begins with the various inspectors who are stationed in the field, in the various flight standards district offices across the nation. There are inspectors who are designated to deal with operations issues, maintenance issues, airworthiness issues, and avionics issues.

In the past, field inspectors have been highly trained individuals who generally came from a background in the industry. But with the tremendous increase in enforcement activity, the FAA has recently hired 600 new inspectors, assigning them to the several specialty areas. The negative side of this rapid expansion in the field inspector force has been that we now have an entire team of new and often less trained individuals enforcing the regulations. Many of these inspectors are younger and, therefore, lack the experience and the operational knowledge possessed by their forbearers.

Only time will tell if these recent hires mellow with age and, further, only that passage of time will enable us to see if the current enforcement attitude and practices soften as well.

Relationship to the NTSB

The National Transportation Safety Board is not connected with or directly related to the FAA. The NTSB functions as a separate safety board and has jurisdiction over virtually all forms of transportation. Its general purpose and reason for existence is to investigate accidents that occur not only in aircraft, but on railroads, in shipping, and to some degree, on the highways.

The NTSB was formed to serve as a sounding board and a safety-related organization that could independently investigate accidents and other, similar failures of the nationwide transportation systems, and recommend changes, improvements, and new standards to improve not only aviation safety but transportation safety in general.

As far as the NTSB directly affects the general aviation industry, it functions as an appellate body in the regulatory enforcement process. Again, Chapter 6 deals with this matter completely and sets forth how the NTSB functions and the procedure and practice necessary to invoke its appellate jurisdiction when some pilot, mechanic, operator, or other certificate holder comes into contact with the FAA enforcement process.

Many people mistakenly believe that the NTSB is a governmental agency that has a superior position to that of the FAA. Such is not the case. The NTSB cannot force the FAA to do anything, but simply recommends. And recommend it does. Often after an accident investigation is complete, the NTSB will make recommendations concerning the issuance of airworthiness directives, a change in operational practices, sometimes a change in air traffic control methods, and the like. But regardless of what the NTSB recommends, it is the function of the FAA to receive that recommendation, analyze it, and act upon it only if the FAA so wishes.

Recently, the NTSB recommended that the FAA conduct a thorough reevaluation of a certain general aviation airplane to see if it met the certification standards for a normal-category airplane. The FAA did conduct those tests and determined that originally the airplane was properly certificated. However, there have been many times when the FAA, after study, has rejected an NTSB recommendation and has refused to act upon it.

The two agencies definitely have intermeshed concerns and, to a slight degree, combined functions. But those functions relate primarily to the enforcement process, and not to the rule-making or policy-making roles of the FAA.

AIRMAN CERTIFICATES AND RATINGS

Five levels of *pilot* (airman) *certificates* are issued by the FAA: *student pilot, recreational pilot, private pilot, commercial pilot,* and *airline transport pilot*. The privileges of each level of certificate vary; holders of one certificate may exercise the privileges of a certificate junior to the one they hold. Let us analyze each in detail.

Student Pilot

The *student pilot certificate* is the first in the five pilot certificates. When I learned to fly in the mid-1960s, the student pilot certificate was a separate license, and a trip to the FAA office to apply for that license resulted in filling out a very official-looking form. At that point it was necessary to receive a *third-class medical certificate* from an aviation medical examiner prior to exercising student pilot privileges.

Today, the process has been somewhat simplified. Because both a private pilot and student pilot are required to possess a third-class medical certificate, which has a duration of 24 calendar months from the date it was issued, the FAA decided it would be very simple to include the form of the student pilot certificate along with the third-class medical certificate. Now, when a beginning aviator receives his physical examination for the third-class medical, and if he successfully passes the same, he is issued both the third-class medical certificate and the student pilot license by the physician who is designated as the *aviation medical examiner*.

No certificate of any kind, including the medical certificate, is required to receive instruction from a flight instructor, nor is there any minimum age to begin flying lessons. I administered dual instruction to my daughter, who at the time was about 7 years old, and logged it as dual instruction in a pilot logbook that I had procured for her. That flight time and that instruction is just as valid for later use as any that she will receive at any later age. (By the way, she did a reasonably decent job of climbs, descents, turns, and straight and level flight by the time she was 8 years old.)

The student pilot certificate and third-class medical certificate are required before the fledgling flyer may fly the aircraft solo. Also prior to solo flight, the certificate must be properly endorsed on its reverse side by the flight instructor, to allow the student to engage in solo flight.

Once that endorsement is made, the student may fly solo, under the direct supervision of the flight instructor, in the local area of the airport in which he is receiving instruction. He may not make cross-country flights beyond that local area until he has received further instruction in the techniques of cross-country flight (including but not limited to weather and flight planning), has had his student certificate endorsed for cross-country flights, and receives an endorsement in his pilot logbook that the instructor has reviewed and discussed with him his preflight planning, for each cross-country flight, to specific destinations.

At no time may a student pilot carry any passenger or operate an aircraft for hire or compensation. Also, the student is limited to solo flights in specific aircraft for which she has received instruction from the flight instructor and an endorsement on her student pilot certificate. One of the fastest ways to end an aviation career is for a student to be caught carrying passengers, flying for compensation or hire, or otherwise violating the strict limitations upon the privileges of a student pilot.

The FAA will usually seek revocation of the student pilot certificate of anyone who engages in such activity, and not merely a suspension of those privileges. The operating limitations of the student pilot are taken seriously, and should never be abridged.

Also, a student pilot may not operate an aircraft on a solo flight in a Class B airspace unless he has received both ground and flight instruction regarding that terminal control area and the flight instruction was received in the specific Class B airspace for which solo flights are to be authorized. He must also have his logbook endorsed within the preceding 90 days, reflecting that instruction and flight instructor authorization for solo flight in that particular airspace.

There is no general restriction upon the types of aircraft in which a student may be authorized to conduct solo flights. It is permissible under the regulations, although rare in practice, that a student pilot could be authorized to conduct solo flights in any aircraft that is certificated for single-pilot operation. There have been stories in the past of particular celebrities who learned to fly in multiengine aircraft, and even turbine equipment. Obviously, that is not the ordinary course of events, although not prohibited.

Recreational Pilot

Effective August 31, 1989, the FAA created a new level of pilot certificate known as the *recreational pilot*. This proposal first came out in 1982 as a result of a petition by certain industry groups. In 1985, the Notice of Proposed Rule Making to create the certificate was issued by the FAA and during the public comment period, more than 2800 comments were received, the great majority of which were in favor of the issuance of this new category of pilot.

The recreational pilot is a middle step between the student pilot, whose activities are severely restricted, and the private pilot, who enjoys a tremendous degree of freedom of operation in the airspace system.

Under the regulation as currently adopted, the recreational pilot is permitted to fly only airplanes and rotorcraft that have single engines of not greater than 180 certified horsepower. In airplane operations, the recreational pilot is limited to aircraft that have fewer than five seats and that do not have retractable landing gear. It appears that the pilot is able to fly an airplane with a constant-speed or otherwise controllable propeller, although few airplanes exist within the horsepower limitations that have such propellers.

The seating limitation is somewhat superfluous because the recreational pilot is permitted to carry only one passenger per flight. She may operate in daylight only, in constant reference to the ground, and may not venture farther than 50 miles from an airport at which she has received instruction. Therefore, cross-country flights of any significance will be virtually impossible for the recreational pilot unless she receives instruction at a chain of different airports, and always flies to and from airports that are no farther than 100 miles apart.

The month of August is special for recreational pilots. This level of certificate was first created in August 1989. In August 1995, while attending the annual convention of the Experimental Aircraft Association at Oshkosh, Wisconsin, the FAA administrator announced proposed new regulations governing recreational pilots. If adopted, and they probably will be by the time this edition of this book is in print, there will be two major changes in how recreational pilots can operate.

The first of these new proposals eliminates the 50-mile cross-country limitation, after a recreational pilot has received dual instruction in cross-country flight from a flight instructor, and gets a logbook endorsement to the effect that the flight instructor considers the recreational pilot competent to make cross-country flights. The second new idea put forth by the FAA will allow the recreational pilot to forgo the requirement to have a third-class medical certificate. The pilot will be able to self-certify her own medical fitness to fly, as has been the case for decades for balloon and glider pilots.

The recreational pilot is also limited to flight by ground reference only, which eliminates VFR-on-top operations. His *visual flight rules* visibility requirement is 3 miles. He also may not fly into any airspace requiring contact with and flight under the purview of air traffic control. This means that the pilot is not permitted to fly from controlled airports, or into Class B or C airspace, or at altitudes above 10,000 feet. If the recreational pilot has not flown within the preceding 180 days, he is required to take a review flight with a flight instructor and have his logbook endorsed to that effect.

As the rules currently stand, unless the new proposals are adopted, if the recreational pilot wishes, he will be able to separately add student pilot privileges to engage in night and cross-country flight, and operations involving contact with air traffic control. In these areas, his operational privileges will be exactly the same as those of a student pilot. He must take instruction in that operation from a flight instructor and receive a logbook endorsement each time that he would like to exercise those privileges; such flights at night, cross-country, or involving air traffic control must not be conducted while carrying any passengers. If the regulations are amended to permit cross-country operations by recreational pilots, then those flights would not be made under the limitations of a student certificate.

There has been much discussion about the propriety of the creation of this intermediate level of pilot certificate. I favor it greatly if properly used. We have come to the place in our pilot training system that learning to fly has become a very expensive procedure. This is not only because airplanes are expensive to operate, but also because today's student who seeks full private pilot privileges is required to demonstrate the ability to control the airplane under instrument conditions, navigate by radio navigational aids, operate into extremely complex controlled airspace, and, if she wishes to be able to fly at night, demonstrate that proficiency and have a minimum of 5 hours of night instruction.

By creating the recreational pilot certificate, we have now enabled a person to do either of two things. He can gain the recreational certificate, and if his desires are simply to fly in that manner in simple airplanes or rotorcraft, and not venture into airspace where contact with air traffic control is required, he can forever enjoy the restrictive privileges of recreational pilot.

But I predict that, more realistically, the recreational pilot certificate will simply be an interlude or a step between the student pilot and the full privileges of the private certificate. It can also be a tremendous opportunity for the pilot to stem the dollar drain of flight training as previously conducted while, at the same time, enjoying some limited amount of freedom from the constant eye of the flight instructor.

Years ago, pilots learned to fly in simpler airplanes and in a world that was not so complicated with positively controlled airspace. Although it is certainly more difficult today to enjoy those operations, it is far from impossible. The recreational pilot is now able to simply pause and take a breather, flying less complicated aircraft in airspace that is not under the domain of air traffic control (ATC) or in which communication with ATC is required.

Hopefully most recreational pilots will seize upon this opportunity and build basic aviator skills to a higher level than that of which a student pilot is capable, before progressing onward to full private pilot privileges. Because the certificate is so new, there is a complete dearth of operational knowledge and statistics upon which to base any accurate forecast of how recreational pilots will perceive and use their certificates. It rests upon the industry and those of us within it to make appropriate use and advantage of the recreational certificate, but the certificate probably will not be seen as an end unto itself, except under one very important exception.

There are many potential pilots who, for some reason or another, cannot pass the physical exam and qualify for a third-class medical certificate. Yet most of them are perfectly safe to fly, if they deem themselves to be so. The percentage of aviation accidents that are caused by pilot incapacitation or other physical problems is so small as to be statistically insignificant. When one compares that very small percentage of physical accidents in airplanes, where medical certificates have always been required, to the problems associated with a pilot's condition in glider and balloon accidents, the percentage is virtually identical. No medical certificate has been required for decades for balloon and glider pilots. What this tells me is that pilots of all categories of aircraft aren't fools. Those who have been able to self-certify their medical conditions all along are not physical wrecks who are waiting to pass out, have heart attacks, or suffer some other incapacitating event at the controls of an aircraft. The instinct for self-preservation is as strong in pilots as in everyone else.

When a pilot self-certifies her medical condition, she is required only to sign a statement that she has no condition which would made it unsafe for her to fly. The pilot does not represent that she meets the requirements for a medical certificate. However, once a pilot has been denied a medical certificate by the FAA, she is not eligible for self-certification. So, we won't have a group of recreational pilots self-certifying who have been refused a medical certificate, just as we don't have them now in gliders or balloons.

Pilots who hold higher levels of certificates—private, commercial, and airline transport pilots—would also be able to self-certify their medical conditions, and fly under the limitations and privileges of a recreational pilot. That change in the rules would allow many such pilots, particularly older or retired aviators, to continue flying on a sport basis, after their more exotic flying days are over.

Private Pilot

The *private pilot certificate* is the first full level of pilot certificate in which the pilot has freedom of operation, and with which certificate, properly rated, he may fly virtually any aircraft. The private pilot may carry passengers in aircraft for which he is rated, and may make unlimited cross-country flights, even internationally. However, the private pilot may not operate an aircraft for any compensation or hire, except that a person who holds the certificate and who has logged in excess of 200 hours may act as an aircraft salesperson and demonstrate his wares to prospective buyers.

The private pilot may also fly an aircraft in furtherance of his personal business, which is a privilege denied both the student and the recreational pilot. Once a person has gained the private pilot certificate, there opens unto him the full gamut of aircraft, providing he achieves the appropriate ratings to operate the same as a pilot in command.

First of all, the private pilot must have an appropriate rating on her certificate for the *category* of aircraft in which she is licensed: *airplane, rotorcraft, glider*, and *lighter-than-air*. Then each category, with the exception of glider, is broken down into *classes*. For *airplanes*, the classes are *single-engine land, multiengine land, single-engine sea*, and *multiengine sea*. For *rotorcraft*, there are only two class ratings: *helicopter* and *gyroplane*. As the current scheme exists, there is no separate class rating for *multiengine helicopters*, although such has been discussed. Lastly, within the *lighter-than-air category* , there are again two class ratings: *airship* and *free balloon*.

There is another proposal by the FAA in the works to change some of the categories to account for new types of aircraft either presently in existence or in the experimental flying stage. A powered-lift category is being discussed to cover the tilt rotor type of aircraft that takes off and lands vertically like a helicopter, but that then flies like an airplane in level flight. Categories for powered and unpowered gliders are under review to require glider pilots to be separately rated for the motor glider and the traditional glider that has to be towed aloft.

Certain aircraft require a *type rating*, which means that in order to exercise the privileges of the private pilot certificate in that particular aircraft, the pilot must be actually certificated for that specific type. The general types of aircraft that require type ratings are large aircraft that have a maximum gross takeoff weight in excess of 12,500 pounds, any turbojet-powered airplane, any helicopter for operations requiring an airline transport pilot certificate, and other aircraft type ratings which the administrator of the FAA may specify through the appropriate type certification procedures when the aircraft receives its type certificate.

The private pilot may also earn an *instrument rating* that authorizes him to engage in instrument flight, operate within the purview of *instrument flight rules*, and fly his aircraft by sole reference to instruments. Instrument ratings are only issued on private and commercial pilot certificates, as they automatically become part of the privileges of the airline transport pilot certificate. Instrument ratings are issued for airplanes and for helicopters. Again, there has been talk of separate instrument ratings for multiengine airplanes and helicopters, although at the present time, there is no separate requirement to be recertified when a single-engine instrument pilot gains his multiengine rating. He must receive instruction in the multiengine airplane for flight under instrument conditions, but there is no separate rating for the operation of multiengine airplanes or multiengine helicopters, by reference to instruments.

Thus, once the private pilot gets his basic certificate and adds to it whatever ratings are required for the particular aircraft he wants to fly, he has virtual and complete freedom of operation throughout the world, with the basic exception that he cannot operate for compensation or hire: he may carry passengers, he may operate into the most densely traveled of airports, he may travel internationally, and do virtually everything else that any pilot may do who possesses a higher level certificate, other than operate for compensation or hire.

While on the subject of compensation, it should be noted that a private pilot may share the operating expenses of a flight with her passengers. Much discussion has been generated about this small section in the regulations, concerning whether the sharing of operating expenses means that the pilot herself must contribute a pro rata share with passengers, or whether the passengers may completely pay for the operating expenses of a flight and the pilot contribute only her services.

The word *share* is fairly simple. I would take the position that a private pilot should pay a pro rata share of those operating expenses, and not merely contribute his services. The ramifications of contributing only the piloting of the aircraft could be stretched to include that the private pilot is being compensated in the sense that he is contributing piloting services, while others pay the total actual expenses of the flight.

Also, a private pilot may act as the pilot in command of an aircraft that is used in a passenger-carrying airlift sponsored by a charitable organization, and for which the passengers make a donation to that charitable organization. There are certain limitations to that limited "commercial" operation by a private pilot. The sponsor of that charitable airlift must notify the flight standards district office that has jurisdiction over the area concerned, at least a week before the flight, and the notification is required to furnish virtually any information that the flight standards district office requests. Furthermore, the flight must be conducted from a public airport, and the private pilot must have logged at least 200 hours of flight time. No aerobatic or formation flights are permitted, and such flights must be made during daylight, under visual flight rules. Lastly, each aircraft used for such a charitable airlift must be certified in the standard category, and must have had a 100-hour inspection in accordance with FAR 91.169.

Commercial Pilot

The *commercial pilot certificate* is the first step required by anyone who wishes to fly for compensation. As we have previously discussed, recreational pilots may not fly in any way connected with business, and although private pilots may operate aircraft in furtherance of their business and, if they meet certain conditions, demonstrate aircraft for sale during their employment as aircraft salespeople, private pilots cannot actually be compensated for services as pilots. However, the commercial pilot may be compensated.

The eligibility requirements are different from those of a private pilot, in that the candidate for a commercial certificate must be at least 18 years old, and must hold a valid *second-class medical certificate*. The medical certificate must be renewed annually for the person whose flying is of a type that requires the exercise of her commercial pilot privileges. However, once the year has passed, the second-class medical certificate continues to be in force for another year for operations that require only private pilot privileges, essentially a third-class medical.

Probably the biggest difference between the eligibility requirements for private and commercial certificates is that the commercial candidate must already possess an instrument rating in the case that he is seeking commercial privileges in airplanes, or whatever commercial certificate that is issued to him will be endorsed with the limitation that prohibits carrying passengers for hire in airplanes on cross-country flights of more than 50 nautical miles, and the pilot will be limited to not being certified to carry passengers for compensation at night under any circumstances. This is a relatively new addition to the commercial requirements.

When I received my commercial certificate in the late 1960s, no instrument rating was required, and I would venture to say that a good number of the applicants for the commercial certificates did not yet have the instrument rating. I did not. Whether safety was increased by requiring all commercial candidates to already possess an instrument rating is a question to which there is no definite answer, and one which can be argued ad infinitum. There is little point in taking time and space to repeat the well-worn arguments both in favor and against requiring commercial candidates to already have an instrument rating, since the issue has been settled for almost a decade, and further ruminations are theoretical only.

The commercial pilot applicant must have at least 250 hours of flight time and no more than 50 hours of that time may be instruction in an acceptable ground trainer. In addition, the commercial candidate in powered aircraft must have at least 100 hours in powered aircraft, and if he is seeking an airplane certificate, at least 50 of those hours must be in airplanes. In addition, he must have 10 hours of flight instruction in a complex airplane, which is defined as one having retractable landing gear, flaps, and some form of controllable pitch propeller.

Further, 50 of those hours must be in dual instruction from a flight instructor, including 10 hours of instrument instruction and 10 hours of instruction in preparation for the commercial flight test. Additionally, 100 hours of pilot-in-command time is required of the commercial candidate, of which 50 hours must be in airplanes and 50 hours must be cross-country flights.

These cross-country flights are defined as at least 50 nautical miles long per leg, and one flight must have landings at three places, one of which is at least 250 nautical miles from the departure point, or 150 nautical miles if conducted in Hawaii. Lastly, as far as experience is concerned, the commercial candidate must have 5 hours of night flying that includes at least 10 takeoffs and landings as the sole manipulator of the controls.

The requirements are significantly different for commercial candidates seeking rotorcraft ratings in either gyroplanes or helicopters, and also for persons seeking glider, airship, or free-balloon ratings.

In FAR 61.139, the regulation sets forth that a commercial pilot may act as pilot in command of an aircraft carrying persons or property for compensation or hire; act as pilot in command of an aircraft for compensation or hire; and lastly, in airships or free balloons, act as flight instructor. There is quite a bit that the regulation in Part 61 does not say.

A person who holds only a commercial pilot certificate may not act as pilot in command of a Part 121 scheduled airline flight, nor may he act as pilot in command of numerous air taxi flights under FAR Part 135.

For instance, in Part 135, an airline transport pilot certificate is required in order to act as pilot in command in passenger-carrying operations in any turbojet airplane, in any airplane having 10 or more passenger seats, or in any multiengine airplane operated by a commuter air carrier.

Other than the foregoing limitations, commercial pilots may serve as pilots in command in Part 135 operations under VFR conditions if they meet additional requirements beyond those set forth for the issuance of the commercial certificate itself. With yet higher qualifications, which are very close to those required for the airline transport pilot certificate, a commercial pilot may be the pilot in command of a Part 135 operation under instrument flight rules in small airplanes that have fewer than 10 passenger seats and are not turbojet powered.

Because of the complexity of the Part 135 regulations governing the use of pilots in command who hold only a commercial certificate, virtually all Part 135 operators today require their pilots to possess an airline transport pilot certificate. The commercial pilot certificate is the only certificate needed to be a corporate pilot, or to do other limited activities for compensation such as taking aerial photographs, carrying parachute jumpers, towing banners or gliders, and making other similar flights that are not passenger-carrying flights, where the passengers are paying for their transportation.

Airline Transport Pilot

The *airline transport pilot* (ATP) certificate is the highest level of pilot certificate currently issued by the FAA. It has been referred to as the "Ph.D. of aviation." The requirements for its issuance are that the candidate be at least 23 years of age, be a high school graduate or its equivalent, possess a *first-class medical certificate* issued within 6 months before the date he applies for a certificate, and comply with all sections of the part of the FARs that apply to the rating sought. Like the other pilot certificates, the airline transport candidate must also be able to read, write, and understand the English language and be of good moral character.

This discussion will limit itself to the airplane rating, and will not address airline transport pilots who desire to be rated for rotorcraft. For the airplane rating, the candidate must have the following aeronautical experience:

- He must hold a commercial pilot certificate, or a foreign airline transport pilot or commercial license without any limitations, or be a pilot in the United States Armed Forces whose military experience qualifies him for a commercial pilot certificate under FAR 61.73.
- He must additionally have 250 hours of flight time as pilot in command (PIC) of an airplane, or as copilot of an airplane performing the functions and duties of pilot in command under the supervision of the PIC or any combination thereof. At least 100 hours of this time must be cross-country, and 25 of it must be nighttime.

- The ATP candidate must have 1500 hours of total flight time as a pilot, including at least 500 hours of cross-country time, 100 hours of nighttime, and 75 hours of actual or simulated instrument time, of which 50 hours must be in actual flight. There is a credit allowed toward some of the nighttime requirements for ATP candidates who have more than 20 night takeoffs and landings, but the credit is only partial. Also, commercial pilots may credit copilot time toward their total requirement of 1500 hours if they serve as copilots in airplanes required to have more than one pilot by the airplane's approved flight manual, airworthiness certificate, or the regulations under which the flights are operated. Individuals serving as flight engineers who are also participating in an approved pilot training program under Part 121, which governs scheduled airlines, may credit some of their flight engineer time. If the applicant for an ATP does not already have an instrument rating, she must perform as part of her oral and flight test those parts required by Part 61 for the issuance of an instrument rating.

An ATP does not have an instrument rating attached to it, as may a private or commercial pilot certificate. The holder of the airline transport pilot certificate is automatically permitted to operate within the IFR system, by sole reference to flight instruments. In other words, the instrument privileges are contained within the basic ATP privileges, and no separate instrument rating is required to operate IFR. FAR 61.171 specifically states that an airline transport pilot has the privileges of the commercial pilot with an instrument rating.

Because the airline transport pilot certificate is issued after a flight test, the ATP privileges will apply only to the particular class of airplane in which the flight test was taken. In many instances, the ATP candidate will present himself for his flight test in a multiengine airplane, if he is receiving his ATP as part of an upgrade to captain for a scheduled airline or Part 135 operator.

Then, his certificate will show that he is an airline transport pilot rated in multiengine airplanes, and perhaps in a particular type of airplane if the flight test is taken in one that requires a type rating. Then, his commercial pilot certificate will be surrendered and an ATP issued, showing the class of airplane in which he has ATP privileges, and perhaps a type rating. The regulation specifically sets forth that he retains the other ratings that he had on the commercial pilot certificate but, of course, may exercise only the privileges of the commercial pilot with respect to those ratings.

To simplify this confusion, suppose that an individual who is a commercial pilot rated in single- and multiengine airplanes and helicopters presents himself for examination for the ATP certificate using a Cessna 310 for the flight test, which he successfully completes. At that point, he would receive an airline transport pilot certificate rated for multiengine airplanes. Then, the certificate would say that he has commercial pilot privileges in single-engine airplanes and helicopters. No mention would be made of an instrument rating, because the instrument rating for airplanes is inherent in the fact that he now has an ATP for multiengine airplanes. So, when operating under circumstances that require only a commercial pilot certificate or a private pilot certificate, the pilot certainly still retains the privileges of an instrument rating that he previously possessed. If he previously possessed a helicopter instrument rating as well, his commercial pilot privileges would remain for helicopters and operation of them under IFR.

In the 1960s, the holder of an ATP certificate was in fact a rare individual. But with the explosive growth in aviation during the 1970s and 1980s, there has been an equal explosion in the number of persons holding airline transport pilot privileges. Because, as discussed above, the ATP is virtually essential for all Part 135 operators and absolutely required for commuter carrier operators, the call for ATPs has increased dramatically.

Today, a person beginning a flying career who intends to make it her livelihood ought to consider the ATP as her eventual goal and continue to progress toward it. In corporate flight departments, ATPs are the norm. Most insurers of large turbine and turbojet equipment look askance at the pilot in command if she does not possess the ATP. So it is no longer seen as an ultimate objective to be achieved by a very few, but as the norm for the truly professional pilot who has made aviation her chosen way to earn a living.

Certificated Flight Instructor

The FAA long ago realized that those individuals who are privileged to be the mentors of new pilots need to be separately certified to exercise the privileges of *flight instructor*. Back in the 1940s, there was a certificate known as a "limited flight instructor," whose holder was an apprentice in the teaching profession. That concept was eliminated several decades ago, and today there is simply one level of flight instructor.

In the late 1960s, drastic changes occurred to the entire scheme of flight instructor certification, duration of the certificates, and the ratings applicable to them. Prior to that change, flight instructor certificates had the same duration as other pilot certificates, and that was until revoked or surrendered. That meant that—just as recreational, private, commercial, and airline transport pilots still enjoy—the flight instructor before the regulatory revisions had lifetime privileges, so long as he did not suffer revocation of his certificate and could continue to pass the applicable medical examinations and receive the required medical certificate.

In the late 1960s, the FAA realized that flight instruction was a constantly changing and evolving activity and with the era of more rapid evolution and change of the airspace requirements, operating techniques, and aircraft designs and types themselves, the FAA came to the decision that flight instructors ought to have some required renewal of their certificates to ensure continued proficiency and competency.

Under the old system, a flight instructor could lay off for 10, 20, or even more years and simply go through the requirements of a certain number of takeoffs and landings to regain pilot-in-command privileges on his commercial certificate, and then, once again, begin instructing students. In fact, this has happened. Look back in history to see that the era of the 1960s was one of dramatic change in aviation.

- Far more IFR operations were being conducted in general aviation airplanes by the end of that decade than at the beginning of it.
- More general aviation airplanes became high-performance aircraft.
- Gone forever, regrettably to some of us, is the multiday cross-country flight in the simple aircraft without radio navigational equipment, and in its place is the retractable-gear, high-speed airplane flown by reference to electronic aids, and quite probably under IFR.

So the FAA realized that the flight instructors certificated after World War II might not be competent to deal with the increasingly complex aviation industry of the 1960s and 1970s. Hence, the new system came into effect and has been with us since.

To be eligible to receive the flight instructor certificate, the applicant must be at least 18 years old, read, write, and converse fluently in English, hold a commercial or airline transport certificate with an aircraft rating appropriate to the flight instructor rating sought, have an instrument rating if applying for an airplane or instrument instructor rating, and pass written, oral and flight tests required for the flight instructor certificate.

Flight instructors must keep records of instructional activity in which they engage. Recreational, private, and commercial pilots must log only the time necessary to meet their respective recent experience requirement. But for the flight instructor, the regulations specify that he must keep records, generally in the form of a pilot logbook, of each flight and the person to whom he has given flight or ground instruction and specify the amount of time and date upon which it was given. The flight instructor must also retain, for a period of 3 years, a record of the name of each person whose logbook or student pilot certificate he has endorsed for solo flight privileges. In addition, he must keep the record for each person for whom he has signed a recommendation for written, flight, or practical test, and must record the result of that test.

Flight instructors are authorized to be the source of virtually all flight instruction taken by individuals who wish to begin flight training, or progress in virtually any manner. Specifically, the flight instructor is authorized to give:

1. Instruction required for any pilot certificate or rating
2. Ground instruction or home study courses required for a pilot certificate or rating
3. Ground and flight instruction to candidates for flight instructor certificates
4. The initial flight instruction required for solo or cross-country flight on the part of a student.
5. A biennial flight review as required by FAR 61.57(a), to be discussed later.

Furthermore, the instructor is authorized to make certain endorsements, including the endorsements on the student pilot certificate for solo flight and solo cross-country flights, the logbook endorsements for a student pilot authorizing solo flights, the review of the preparation of preflight planning and authorization for a student's solo cross-country flights, the logbook endorsements certifying that the pilot or flight instructor is prepared for certain further written or flight tests, and lastly the logbook endorsement of pilots to whom the instructor gives biennial flight reviews.

Flight instructors have certain regulatory limitations upon their activities. For instance, they may not conduct more than 8 hours of flight instruction in any period of 24 consecutive hours. Importantly, a flight instructor may not give flight instruction required in a multiengine airplane or helicopter unless, prior to that instruction, the instructor has 5 hours of experience as pilot in command in the specific make and model of that multiengine airplane or helicopter. As can easily be seen, this is beyond the simple requirement for the carriage of passengers by other pilots who are required only to be rated for a particular category of aircraft and have three takeoffs and landings in a particular type within the preceding 90 days.

As previously stated, flight instructor certificates must now be renewed. The basic period of validity of a flight instructor certificate is 24 calendar months from the time it is first issued or last renewed. An instructor may return for reexamination within the 24-month period of time and take the practical test again.

Obviously, that is not a preferred method for most holders of instructor certificates, and the local FSDO offices of the FAA have other discretionary forms of satisfying themselves that the instructor is competent and can be permitted to enjoy another 2 years of certification without taking the practical test. The methods that a flight instructor can use to renew his certificate without retaking the practical test are:

- If the FAA office is satisfied that his record of instruction shows that he is competent
- If he has a satisfactory record as a company check pilot, chief flight instructor, pilot in command of a scheduled airline, or other activity involving the regular evaluation of pilots, and passes any oral test that the FAA considers necessary to determine the extent of his knowledge of current pilot training and certification requirements
- If he successfully completes, within 90 days before the application to renew his certificate, an approved refresher course consisting of not less than 24 hours of either ground or flight instruction

If the instructor does not renew his certificate in accordance with the above requirements within 24 calendar months of the time it is issued, it expires and he no longer enjoys the privileges. Calendar months is a method of time measurement used quite consistently throughout the Federal Aviation Regulations. It simply means that a period of time expires on the last day of the twenty-fourth month following the day upon which it first occurred.

As an example, if one receives a flight instructor certificate on April 1, 1989, it is valid until April 30, 1991. Hence, 24 calendar months can be almost 25 actual months. Likewise, in areas of the regulations where things are valid for 12 calendar months, such as annual inspections on airplanes and second-class medical certificates, if one receives a medical certificate, or if the airplane is properly signed off for an annual inspection on April 1 of any particular year, that medical certificate or annual inspection is valid until April 30 of the following year.

Once a flight instructor has allowed her certificate to expire, she has only one choice to reinstate it—retake and successfully pass the practical test prescribed for original issuance. Therefore, most flight instructors who intend to use their certificates, even sporadically, go to great lengths not to suffer expiration of their privileges. Most of these individuals will go through the refresher course, which is one of the alternate methods of renewal of the instructor certificate.

Again, it must be remembered that the entire concept behind the unique fact that flight instructor certificates are valid for only a limited period of time is the realization that aviation is changing, and doing so quickly. The refresher course will serve to keep instructors up to date on new training requirements, new certification requirements, and generally new operating procedures within the aviation system. But, once expired, a flight instructor certificate must be reinstated, and that procedure is not painless.

CURRENCY REQUIREMENTS AND FLIGHT REVIEWS

The regulations have, virtually since their inception, had to deal with the problem of pilots who have not flown as pilots in command for a period of time. They set forth standards concerning how often, and in what operations, a pilot must fly in order to continue to operate as pilot in command. It would go without a great deal of opposition that someone who had not flown for 10 years ought not to be able to load a King Air full of passengers and launch into the great unknown. Therefore, the regulations dealing with general aviation pilots have set forth currency standards, which relate to recent experience requirements. These requirements are contained in section 61.57 of the Federal Aviation Regulations.

First of all, a pilot must undergo a biennial flight review, every 24 calendar months. One of the perceived omissions in the flight review requirement is that it must be taken only in an aircraft for which the particular pilot is rated. The meaning of this is subtle, but upon inspection, becomes quite glaring.

For instance, I possess a commercial certificate that contains ratings for both single- and multiengine airplanes, gliders, and helicopters. Anyone who has any real exposure to airplanes, gliders, and helicopters knows that there is a world of difference between them—aerodynamically, physically, and most important, as considered here, operationally. In fact, there seems to be no real correlation or transfer of skills, from the mechanical aspect of flying, to make any real difference, whether a pilot first learns to fly an airplane and then a helicopter, or conducts the initial training in a helicopter and, after being certified in it, goes on to become rated for airplanes. There is a substantial carryover of skills from gliders to airplanes, except for the obvious differences inherent in the fact the airplanes have engines and gliders do not.

Yet, as the regulations now stand, a person who is certificated such as I am may undergo his flight review in a Cessna 150, and have the flight review portion of his recent flight experience requirement satisfied for single- and multiengine airplanes, gliders, and helicopters.

There has been much discussion and commiseration about this oversight in the regulations, and I expect it to be addressed and revised shortly. Perhaps by the time this book is in print, we will see flight reviews required in each category and class of aircraft for which the particular pilot is rated, if the pilot wishes to continue the ability to exercise the privileges of his or her certificate in each category and class of aircraft.

In any event, it is absolutely imperative that the person designated to administer the flight review make a logbook entry in the pilot's log, certifying satisfactory accomplishment of the review.

The flight review is generally given by a flight instructor and, up until recently, the total content of the flight review has been left up to the discretion of the flight instructor administering it. But now, it is required that the review consist of at least 1 hour of flight instruction and 1 hour of ground instruction, at a bare minimum. If the flight review is taken in a glider, by a rated glider pilot, 1 hour of ground instruction is required, but the 1 hour of flight time is relaxed if the pilot undergoing the review makes at least three glider flights in the conduct of the review. Thankfully, then, gone are the days when an instructor could simply go around the traffic pattern once or twice with someone he knows and endorse his log as successfully completing the flight review.

There are other ways of accomplishing the flight review requirement, other than actually taking and successfully accomplishing it. If a pilot satisfactorily completes a pilot proficiency check conducted by the FAA, or an approved pilot check airman for a branch of the United States military, or successfully completes the flight test for the issuance of a new certificate or rating, those acts, and each of them, suffices as a flight review for the ensuing 12 or 24 calendar months as required for the particular pilot involved.

Then, except for operations requiring an airline transport pilot certificate, or other operations conducted under Part 135 of the regulations, there is a general experience requirement. This requirement states that, in order to serve as pilot in command of an aircraft that requires only one pilot, or to be a crew member in an aircraft requiring more than one pilot, each person must perform three takeoffs and landings within the preceding 90 days, as the sole manipulator of the controls of the aircraft. These three takeoffs and landings must have been accomplished in an aircraft of the same category and class as the aircraft which is now to be flown and in which the pilot is to be considered as having satisfied the general experience requirement. If the aircraft is one in which a type rating is required, the three takeoffs and landings in the preceding 90 days must be accomplished in that particular type of aircraft.

One subtlety in this area of the regulations concerns tailwheel airplanes. When I learned to fly, we called them "conventional gear," but they certainly are no longer the convention within general aviation. Now the term *tailwheel airplane* is used to describe what was once the commonplace design of light, single-engine aircraft.

In any event, when a pilot wishes to satisfy the general experience requirement for a tailwheel airplane, the three takeoffs and landings within the preceding

90 days must each be made to a full stop—touch-and-go operations do not count. This is because of the unique handling qualities of tailwheel airplanes during the initial part of the takeoff roll, and the fact that anyone who knows anything about flying them realizes that you do not stop flying one until the engine is stopped and the plane is tied and chocked, or safely put in the hangar. As I have over 1500 hours in tailwheel airplanes, from the Aeronca Champ to the DC-3, that is a fact to which I can personally attest.

In order to exercise the privileges of an instrument rating, and that means file and fly in the IFR system or in actual IFR weather, a pilot has to meet certain IFR experience requirements in addition to the flight review and general experience requirements.

First of all, instrument pilots can retain their instrument privileges by flying at least 6 hours of instrument time, which may be either actual or simulated, every 6 calendar months. This must include at least six instrument approaches. If the pilot does not fly that much instrument time (and by the way, 3 hours of the 6 must be in the category of aircraft involved), the pilot may not exercise instrument privileges until recent experience is regained.

Next, the instrument pilot has an additional 6 months in which to gain the required 6 hours of experience and six instrument approaches. Once 12 calendar months have elapsed since the pilot last met the recent IFR experience requirement, the holder of an instrument rating then has only one way to regain those privileges, which is to successfully take and pass an instrument competency check. The instrument competency check must be given in the category of aircraft involved, and may be given by an FAA inspector, a member of any branch of the military authorized to conduct the flight test, an FAA-approved check pilot, or a certificated flight instructor. It is possible to conduct some or all of the instrument competency check in a simulator, if the simulator is approved for that purpose.

Other sections of the regulations govern the pilot-in-command requirements for proficiency in aircraft requiring more than one pilot, but they will not be addressed here, as this work is geared toward the general aviation pilot, who will probably never be a pilot in command of an aircraft requiring more than one pilot. Should that event occur, the reader is referred to section 61.58 of the Federal Aviation Regulations for those required proficiency checks.

MAINTENANCE CERTIFICATES

Since early in the scheme of the regulation of aviation by the FAA, maintenance personnel have been required to be certificated in order to perform various tasks in keeping aircraft under repair and airworthy. The FAA issues certificates for individual mechanics, to perform maintenance work on airframes and, separately, aircraft power plants.

Also, certain licensed mechanics are granted *inspection authorization* (IA) by the FAA to perform required inspections, and since that field of endeavor most directly affects general aviation, the services and approval of a mechanic who possesses an inspection authorization are required for the annual inspection that must be performed on all aircraft operating under Part 91 of the Federal Aviation Regulations.

In addition to certificating individuals, the FAA issues *approved repair station certificates* to operators who wish to gain a blanket approval for work done by their particular facilities. We shall not delve into the requirements for certification as an approved repair station. The amount of equipment that must be possessed and available for use by the approved repair station varies dramatically by the particular maintenance tasks for which the facility seeks FAA approval. In any event, even approved repair stations are required to use certificated mechanics for certain operations, and must have at least one such individual on staff who possesses an inspection authorization.

Airframe Mechanic

In order to gain certification as an *airframe mechanic,* an individual must serve a prescribed period of apprenticeship working under a licensed airframe mechanic. This period of learning varies, depending upon whether the candidate is seeking certification for only the airframe license or is serving a twofold apprenticeship, after which he or she wishes to gain both airframe and power plant certification. These periods of time vary from 1 to 2 years.

Several educational institutions are approved by the FAA as training facilities for the purpose of providing the necessary instruction for persons to earn an airframe certificate, power plant certificate, or both. If an individual attends such a school, successful graduation from it generally satisfies the apprenticeship requirement.

While working under a licensed mechanic, or attending the approved training school, the candidate must pass a rigorous written examination for each maintenance certificate that he or she wishes to receive. After passing the written test, and upon completion of either the apprenticeship period or the approved training curriculum from a school, the candidate then must pass a practical examination, administered by either an FAA maintenance inspector or a designated examiner, that tests actual ability to perform certain repairs.

The airframe mechanic is licensed to maintain and repair the portions of the aircraft that do not come under the definition of those tasks for which a power plant certification is required. The airframe mechanic repairs and maintains virtually all the systems of the aircraft except those of the power plant and systems directly related to the power plant.

Power Plant Mechanic

An individual who possesses a *power plant mechanic's certificate* is permitted to maintain and overhaul power plants. In days gone by, the power plant mechanic was referred to as an engine mechanic, and a person who possessed both airframe and engine licenses was referred to as an A&E mechanic. Several years ago this antiquated terminology was changed, and in place of the word *engine,* the FAA began using the term *power plant* to designate systems that power aircraft. As jet engines became more prevalent in aviation, it was determined that the use of the term *engine* to describe various types of power plants was obsolete. Hence, we now refer to the power-supplying system as a power plant rather than an engine, and we refer to those who are licensed to maintain them as power plant mechanics rather than engine mechanics. Therefore, today, the mechanic who holds both airframe and power plant certification is referred to most commonly in the industry as an A&P mechanic rather than the former nomenclature of A&E.

Virtually all maintenance activities not specified in Part 43 of the FAR as allowed to be performed by aircraft owners or operators who are pilots must be performed by licensed mechanics. It should come as no great surprise that the very foundation of the physical safety of all aircraft results from how well they are maintained once they leave the manufacturing facility. For this reason, the FAA has, since the very beginning of the regulation of aviation, prescribed maintenance standards and practice, and regulated those who perform maintenance functions. Licensed aviation mechanics, most of whom possess both airframe and power plant certificates, are true professionals and possess training, experience, and skill levels commensurate with the heavy responsibilities attendant to their jobs.

Those items of minor maintenance that unlicensed personnel are permitted to perform and approve are dealt with more fully in Chapter 3. Those minor maintenance jobs can be best categorized as preventive maintenance, and not as any significant repair of the aircraft or any of its systems. The latter always requires the services of licensed maintenance personnel who, if they do not actually perform the work, supervise the work of apprentices serving under them, and add their signature, approval, and attendant liability to the work that is done.

Inspection Authorization

An individual can apply to the FAA for designation for *inspection authorization* after working as a licensed mechanic for a minimum of 3 years. In order to gain inspection authorization, the mechanic must successfully pass another written examination and oral quizzing. The function of the authorized inspector is just what the term implies; it is he who is given the authority by the FAA to conduct required inspections of an aircraft in addition to his continuing approval, pursuant to his mechanic certificates, to do repairs.

The regulatory scheme recognizes that the mechanic with inspection authorization is required to possess a higher degree of skill and experience than is the person who may only repair and maintain aircraft, without the ability to conduct the required inspections. In order to qualify to act as an authorized inspector, the candidate is required to have a greater degree of familiarity with different types of aircraft and is virtually required, by the nature of the exams and the work, to possess a working library of airworthiness directives and service bulletins affecting the various types of aircraft that will come into his shop.

The inspector plays a crucial role in aviation safety. While no licensed mechanic is assumed to perform substandard work, or to permit such by those apprentices working under him, the role of the authorized inspector goes far beyond simply performing needed maintenance in a workmanlike manner. When an aircraft comes into the shop for an annual inspection, and the individual with inspection authorization approves it for return to service and signs its airframe and power plant logbooks indicating that both the aircraft and its power plants are airworthy, he is putting his reputation, certificate, and fortune on the line for not only the maintenance and repairs that happened to have been performed during the course of that inspection, but virtually for the entire history of the aircraft.

It is the mechanic with inspection authorization who is responsible to search the applicable airworthiness directives for each aircraft on which he performs an annual inspection, and determine that all applicable directives have been performed or in some other manner complied with before he certifies the aircraft as airworthy for another year. It is also the responsibility of the inspector to determine that all previous repairs or modifications made to the aircraft were done in a workmanlike and approved manner, and that the airworthiness of the aircraft was not affected adversely.

The role of those with inspection authorization is indeed weighty. They may well, in the course of certifying an aircraft to be airworthy, be biting the bullet for work of unknown quality, done at unknown times, and by unknown persons, at times in the past that cannot be determined. A great deal of discretion and judgment is required of inspectors, and those of us who are pilots constantly owe them all a debt of gratitude and a tip of our hats for the excellent job that is done throughout the industry in keeping the general aviation fleet in safe operating condition.

3

Aircraft Ownership

ONCE MOST INDIVIDUALS OBTAIN A PILOT CERTIFICATE, IT IS NOT LONG until they begin to contemplate the day when they may own their own airplane. Many questions arise during the formulation of the desire to own an airplane. This chapter deals first with the purely mechanical aspects of aircraft purchase and how those aspects need to be viewed in light of certain legal principles. Once that chore is completed, an examination follows of the various ways that aircraft might be owned by more than one person and some of the positive and negative consequences of joint or co-ownership.

ACQUISITION OF AN AIRCRAFT

The first step in the purchase of any aircraft is to decide the basic performance characteristics and parameters that are needed to fly a particular mission or missions contemplated by the prospective purchaser. The first rule in this area is to be honest with yourself. If your average need is for an airplane to carry one or two people for distances of no more than 100 to 200 miles (on a routine basis), do you really need a six-place, retractable-gear airplane and the associated costs, for both acquisition and maintenance? All of us who own an airplane dream of more speed, more capacity, and higher altitudes. But realism and pragmatism must set in eventually.

I firmly believe that too many people are missing out on the true joy of aircraft ownership because they are not willing to purchase the aircraft that meets their routine needs within an affordable price range. It makes far more sense to own a four-place, or even a two-place, fixed-gear, simple airplane if that will satisfy 90 percent of one's use, and then rent a larger aircraft for the very few flights per year that dictate either more capacity or more speed.

The first question that must be answered is whether you want an airplane for sport and recreational flying, or whether your needs are more of the true transportation variety. If you want an airplane for weekend pleasure hops to nowhere in particular, there is a host of airplanes that meet that need. Do you want to fly into Class B airspace? If you do, you'll need more avionics equipment than you will for flying in and out of rural airports. Just keep in mind that when it comes to airplanes, every additional piece of equipment, and every increase in operational flexibility, comes at an exponential increase in cost.

If your real desire is to fly around the local area in nice weather, and if you aren't flying from a controlled field, there are some classic airplanes that will get you aloft for far less, in acquisition costs, than the price of an average new car. These old classics, if you buy one that has had good care, can be operated by just about anyone without taking out a second mortgage to pay for it. Annual inspections can be as little as $300 in many parts of the country, and their little engines sip fuel at a miserly rate.

When you get into a panel full of radios and black boxes, be ready for the cost of maintaining not only the electronic gadgets themselves, but the airplane's electrical system that powers them. One mechanic friend of mine advises folks that once you add an electrical system to an airplane, figure that your maintenance costs have just doubled. But if you live in a metropolitan area, full of Class B or C airspace, you'll have to bite the bullet and accept the costs, both in terms of purchase price and maintenance, that go along with a more capable airplane.

In Chapter 4, we'll devote our attention to the subject of amateur built aircraft, which are commonly called *homebuilts*. For the person with the right combination of free time, mechanical aptitude, and space in which to build an airplane, whether it be from a kit or a set of plans, homebuilding can be the answer to getting one's own airplane.

In any event, once through the initial decision-making process, a prospective purchaser needs to identify those particular types of aircraft that fit both need and budget. At this point, the services of an aircraft broker might become quite useful, if not necessary. Neither the newer pilot nor the one who is simply less experienced and has not had the opportunity to fly many different types of airplanes is in a position to justly include or eliminate all the potential types that might serve his or her needs. Therefore, some assistance is necessary in this job.

Often pilots will seek the advice of their flight instructors, other apparently more experienced pilots, magazine articles, and a host of other sources of information to attempt to narrow the great field of general aviation airplanes down to a more manageable number of aircraft types from which to eventually choose. A prospective purchaser will do well to gain as much information possible from all available sources. A good source, and often the best source, is a knowledgeable and honest aircraft broker.

Because brokers work in what is in essence a national, if not continental, marketplace, it generally does not make a lot of difference to brokers if they have a particular line on or already own the aircraft or various types of aircraft in which a prospect might be interested. Once the broker assists the prospective buyer in selecting the type or types of aircraft in which that buyer is or might be interested, the broker can quickly obtain information about particular specimens for sale from several nationwide sources.

Prepurchase Inspection

Assuming that our particular buyer is interested in a used aircraft as opposed to new, one of the first criteria ought to be the successful completion of a *prepurchase inspection*. Once a particular aircraft is identified and the prospective buyer has become interested in it—enough so that she is now willing to say to herself and to the possible seller that she will buy the aircraft if it is in a condition, such that it is, as represented to her—our buyer ought next to arrange for an inspection. The inspection needs to be done by a maintenance facility or mechanic in whom the buyer has faith and confidence, and naturally who is not in any way connected with or particularly loyal to the seller.

Although a prepurchase inspection by its very nature must fall short of a more complete inspection of the aircraft, such as is done on an annual basis to meet Part 91 requirements, a good prepurchase inspection can be thorough, especially for the less complex airplane.

A skillful and experienced mechanic can, in a few hours, delve deeply into the innards of virtually any fixed-gear or fixed-propeller airplane. Specifically, the mechanic can get in deep enough to identify any skeletons that might be hiding in the various nooks and crannies of the aircraft. As the complexity of the airplane under consideration increases, so does the amount of time and necessarily the cost required for an equally thorough prepurchase inspection. But even several hundred dollars spent at this stage can pay great dividends down the road, should unexpected problems be identified before the purchase is made. Then a decision can be made either not to go forward or to renegotiate the proposed financial terms to compensate for any repairs that might be immediately, or in the very near future, necessary.

The buyer needs to make sure that the prepurchase inspection is performed by a mechanic of the buyer's choosing, not the seller's. If the aircraft must be flown by the seller to that facility, it is not at all unusual for the seller to request reimbursement for at least the fuel and oil consumed in the flight, and the buyer should be willing to offer such. If the seller shows any hesitancy in allowing an airplane to be put to a good prepurchase inspection, my personal practice is to immediately shy away from the airplane and thank its owner for his time but firmly inform him that I have no intention of pursuing any further business with him.

For every rule, exceptions must exist. If the right opportunity presents itself for the experienced and sophisticated buyer with sufficient financial reserves, virtually any airplane is worth buying in less than ideal condition if the price is right. Again, the services of a trustworthy broker or other adviser can be indispensable in this area. Many people have knowingly purchased aircraft with engines that are at or beyond the manufacturer's recommended time for overhaul, knowing that they will immediately have to sink thousands (if not tens of thousands) of dollars into significant overhaul or other major repairs.

With the appropriate degree of knowledge of the condition of the airplane and the attendant financial costs to bring it into first-class condition, and with the desire to keep the airplane and fly it the hundreds of hours that are necessary to successfully amortize such an investment, the buyer will often find the purchase of aircraft in that condition to be a satisfying experience after the renovations or restorations are complete.

A few years ago, I was interested in buying an old Taylorcraft. Why I had this desire is incapable of precise definition. Maybe it was because my very first airplane was a BC-12D Taylorcraft, which I owned back in 1965 and 1966. I flew that airplane, which was a few months older than I, for about 200 hours, building time for my private license, and later for my commercial certificate. It took me from central Ohio to Florida, upstate New York, and all over the midwest.

I found out about one that was for sale at a rural airport near my home. From the asking price, I knew that it would take some work to put it in pristine shape. When I first looked at it, with the owner selling hard to convince me that it was a decent airplane, it looked as though he might have been right. I'm no stranger to mechanical things, and have been around airplanes all my adult life. Plus, there isn't anything very complicated about such a simple airplane.

But I let the better part of judgment control, and I called a friend who runs a maintenance facility where he works on many classic aircraft. He went with me to do a prepurchase inspection. After about 2 hours of looking, prodding, and frowning, he called me aside to discuss his findings, which were recorded on two sheets of a typical yellow legal pad. He had three lists.

The first he called "repairs needed before further flight." This list included such things as a landing gear leg so corroded that it might have failed in any but the smoothest of landings, and the fact that one of the four cylinders of the engine had absolutely no compression. His next list contained discrepancies that would have to be fixed before he would sign the aircraft off for its next annual inspection. The last list contained optional items to be corrected to put the airplane into the condition he knew that I would want. This inspection, and the results which my friend, Luke, communicated to me, was about an airplane that some mechanic had passed for an annual just a few months before, and that was being flown fairly regularly.

The sum total of the costs of making the repairs on the three lists exceeded the asking price for the Taylorcraft. Needless to say, I didn't buy the airplane. The moral of this story is simple—get a prepurchase inspection done on any airplane you are considering for purchase, regardless of how nice it looks. And get the inspection done by a mechanic or a shop familiar with the type of airplane.

In the last analysis, there is no definite statement of the right way or wrong way to go about the decision of which aircraft to purchase. The options are many and the pitfalls are several. One of the leading investment brokerage houses in this country uses the saying "invest your time before you invest your money." There is no better cliché to succinctly describe the appropriate way to go about the initial selection of an aircraft to acquire.

Contract of Sale

As will shortly be seen, the FAA requires very little actual paperwork to register the sale of an aircraft and accomplish the transfer of the title to it as shown by the chain of title and other documents recorded at the FAA's Aircraft Registry. But, as with most other things, simply meeting the bare minimum requirements is seldom the course of action that ends up with the best results.

Rather, an expressed written *contract of sale* should be prepared and entered into for the purchase of an aircraft. In a manner similar to real estate, many legal questions can arise concerning the ownership of airplanes that cannot be fully and completely disclosed by a search of the records at the Aircraft Registry. A contract of sale not only needs to cover all the representations that have been made by the seller concerning the physical condition of the airplane, the time on its engines, the airframe, and the like, but should incorporate them into written promises in the contract.

The contract needs to clearly set forth all the terms of the sale, which include not only the purchase price, but the point at which the aircraft is to be delivered, who is to pay any sales or use tax that might be levied by any particular state taxing authorities, and the status of all required airworthiness directives and mandatory periodic inspections, such as those for the pitot-static system and transponder. The services of a knowledgeable aviation attorney can be well worth the small financial investment to get a contract of sale drawn that will protect the buyer from the unexpected and unknown.

Likewise, a good contract is also in the best interests of the seller. The seller can make whatever disclaimers that he wishes concerning not only specifics, but also all the implied representations or implied warranties that the law might thrust upon him if he is not careful to exclude them.

The law contains many arbitrary rules to fill in gaps where parties to a transaction fail to be specific. Such things as the time and manner of payment, the place of payment, the condition of the aircraft itself, and similar important terms can be governed in what, in essence, are two ways. First, the parties can negotiate and agree on such matters and set forth their agreement. Second, if they choose not to do so, or simply fail in that regard, the law will come forth with its arbitrary rules to settle future disputes. In my years of practice, I have seen few people totally satisfied with the mechanisms that the law must impose when parties fail to set forth their agreement or even fail to reach agreement on important, but perhaps uncontemplated, provisions.

Title Search

The FAA has established, pursuant to federal law, a department at the Mike Monroney Aeronautical Center in Oklahoma City that is known as the Aircraft Registry. For many reasons, not the least of which is the fact that airplanes are probably the most mobile pieces of equipment civilization has yet devised, it became apparent decades ago that there needed to be a central repository for the information that determines the ownership of aircraft as well as the liens or other encumbrances which might be in existence against it, and that such information needed to be available to be public. Therefore, the Aircraft Registry has become that repository.

It is beyond the scope of this book to delve deeply into the court decisions concerning the resolution of competing claims to the ownership of aircraft. When we refer to *title* to an airplane, we are not referring to a single piece of paper as we normally do when the word *title* is used with respect to a motor vehicle. Virtually all states have decided that the best way to solve such problems concerning automobiles, boats, outboard motors, trailers, and motorcycles is for a state agency to be responsible to issue a document, normally referred to as a certificate of title, setting forth the identity of the owner of that specific vehicle. In many states, including my own, such documentation is viewed by the law as conclusive proof of ownership of the vehicle referred to on the certificate of title.

However, the title to aircraft is not even similar. Those readers who are conversant with the subject of real estate law, are best able to draw an analogy between that and the law of aircraft ownership: Just as a deed to a particular parcel of real estate does not, in and of itself, establish that the grantee whose name appears on the deed is in fact the owner of the real property, neither is the person whose name appears on a certificate of registration of a particular aircraft necessarily the owner of it.

There are times when the Aircraft Registry may refuse to accept a bill of sale and register the aircraft in the name of the person who just paid for it. If this happens, the buyer is often in for a long and expensive legal process in trying to clear the title so that the FAA will consider the aircraft to be the buyer's property, and issue a certificate of registration. I recently finished such a problem for a client who bought a helicopter fuselage, intending to rebuild it into an airworthy machine.

The helicopter had a chain of title that was nightmarish, with missing conveyances from certain corporations to others, modifications made to the airframe that were not documented with recorded Form 337s or Supplemental Type Certificates, and an elusive prior owner who we initially thought was living in South America. For the better part of a year, the seller had my client's money, my client had no usable aircraft, and the fuselage was sitting at a repair facility awaiting the solution of all these problems before any more money could be sunk into beginning the rebuild. A good title search and examination of it by an aviation attorney could have prevented all this grief for the buyer.

Many factors, such as *unreleased security agreements* (formerly referred to as *chattel mortgages*), *liens by consent* to repair facilities, *liens* filed by repair facilities from certain states, and *former ownership* by groups of individuals who have not properly conveyed their title to persons previously in the chain of ownership, all can create clouds upon the free and marketable title to an aircraft.

The only solution to this morass is for a prospective buyer to have undertaken a good and thorough *title search* of the records at the Aircraft Registry. Unlike in real estate, where the records that affect title to the property exist in the offices of some county official at the county seat of the county in which the property is located, there is no practical way for the records affecting aircraft ownership to be situated in any place other than one national and central location. Again, the FAA Aircraft Registry in Oklahoma City is that place.

Because it is obviously impractical for persons in most of the country to journey to Oklahoma City, Oklahoma, to conduct the record search personally, and further because a good number of folks would not know what they are examining or the legal implications of the various documents, several title search services exist in the Oklahoma City area that provide, for a relatively nominal fee, searches of the documents at the Aircraft Registry, and then issue a report to the prospective buyer or other interested person of the results of that inspection.

To purchase an aircraft without a title search borders on the foolhardy. A recommendation can be had toward the identity of the various searching companies from many sources. Again, reputable aircraft brokers will normally provide such a title search as part of the services in the transaction. If the prospective buyer is intending to finance the purchase of an aircraft, it is almost certain that any institutional lender will demand a title search and report prior to funding any loan. Those lenders that frequently do business financing aircraft are another source of recommendation and identification of the various title search companies. If a prospective buyer intends to pay cash for the airplane without the need to institutionally finance it, organizations such as the Aircraft Owners and Pilots Association and similar groups often either have title-searching companies allied with them, or at the least, can serve as a source of recommendation.

However, simply getting the search performed and receiving a written report of its outcome is seldom sufficient. Just as the average person cannot adequately interpret a report concerning the records at the county seat relating to the ownership of real property, neither can she reasonably assure herself that the documents reported in the title search indicate that the aircraft is properly owned by the potential seller and that when the purchase price is paid, the buyer will obtain the ownership of the aircraft free and clear of all encumbrances. Therefore, the title search report needs to be examined by a competent aviation attorney just the same as the report of a real estate title search needs to be examined by a real estate lawyer.

With the price of aircraft often equaling or exceeding the average price paid for parcels of real estate, the peace of mind and risk elimination achieved are well worth the small fee generally charged by competent counsel to review and render an opinion on the title. Those of us who have been involved in aviation for any time are readily aware of the problems of theft of not only aircraft, but component parts and accessories. But theft alone is not the only problem that a buyer can face if he is not careful and does not enter into the transaction with his eyes open—and with competent and thorough advice.

I would never buy an aircraft without a title search, and I never advise clients to do so either. It behooves all airplane buyers to make sure they are not buying the Brooklyn Bridge, and the costs involved in doing so are frankly insignificant when compared with the price of the aircraft and the potential costs of solving undiscovered title problems that come to the fore after the purchase is complete, and the seller has removed himself either to parts unknown or perhaps halfway across our continent.

Aircraft Title Insurance

In many situations it is wise to purchase a policy that insures that the buyer of an aircraft will, upon payment of the purchase price and the receipt of a properly executed bill of sale, acquire good title to the aircraft that he is purchasing. Such a policy is known as *aircraft title insurance*.

Again, an analogy can be drawn to the real estate industry. For several years, it has been more and more common for purchasers of real estate to purchase title insurance, issued by a title insurance company, that insures the buyer of the property will gain good title to it once he pays the contract price to the seller. That concept has now carried over into the aviation industry as well.

If there are technical encumbrances upon the title to the aircraft that cannot be removed, it is possible to purchase title insurance that would then insure that the owner will not have his title divested by the operation of any of those existing "clouds." When this situation arises, some negotiation will need to occur between the prospective purchaser and the aircraft title insurance company. These types of coverages are not issued automatically, and title insurance companies quite frequently shy away from insuring around known encumbrances or clouds upon the title.

For those buyers who wish additional peace of mind, it is not unrealistic or uncommon to purchase title insurance even when there are no known or apparent problems with the state of the title to the aircraft in question. Just as title insurance for real estate gives the owner a level of comfort, the same can be achieved in the aircraft sales industry when a policy of title insurance has been issued.

Because aircraft title insurance is a relatively new form of coverage, its use is not particularly widespread as of this writing. But just as lenders who deal in the real estate industry insisted upon title insurance to protect their interests—and by so doing drove the real estate title insurance industry into a position of prominence—it is not at all unlikely to foresee that the same might well happen to the aviation industry. With the ever-increasing cost of aircraft, it becomes more important to lenders to ascertain, to the highest possible degree, that the aircraft upon which they are making a large loan will not be subject to the claims of persons other than the immediate borrower. It might be the lending industry that will make the demand for aircraft title insurance as a part of the conditions under which it will make the loan of the purchase price, or a portion thereof.

Only time will tell if aircraft title insurance will become as commonplace as real estate title insurance. It certainly makes sense to expend the small, one-time premium to insure the title to the very aircraft being purchased when several times that amount will be spent each year for liability and property damage insurance to permit the owner to operate it.

The potential purchaser of an aircraft ought to seek information from the various title insurance companies, which are often colocated and represented by the title search companies, as to the conditions of coverage and the premiums charged. The presence of aircraft title insurance in a transaction does cast a net of safety around the owner's present and future interests that cannot be equaled in any other manner.

Bill of Sale

In order for the Aircraft Registry to have a document that indicates that the ownership interest has been transferred from one owner to the next, a piece of paper is required. Because airplanes are in the broad category of personal property, it has been the general custom within the American legal system to use a *bill of sale* as the document to accomplish transfer of title of that category of property, as opposed to the various deeds that are used to transfer real property.

The official FAA bill of sale form is displayed in Figure 3-1. Note that it is a rather simplistic document, about which comment was earlier made. This bill of sale should be completed and used as the document that is to be sent to the Aircraft Registry, along with the other documents that are explained in a few moments, to be a part of the package of paperwork that will result in the transfer of title from the former owner to the new, the recording of any financing documents if the purchase is financed, and the registration of the aircraft in the name of the new owner.

It is important that the seller's signature be properly displayed on the bill of sale. If the seller is a group of individuals that is small in number (generally four or fewer), I recommend that all members of that group sign as sellers. If the selling entity is a corporation, the bill of sale needs to be signed by the appropriate corporate officer as required by the law of the state in which the transaction is occurring, and the buyer should obtain an appropriate resolution of the board of directors of the corporation authorizing that particular officer to sell the aircraft, and sign on behalf of the corporation all documents necessary to consummate the sale. If the selling entity is a partnership, a general or managing partner should sign, and the buyer should similarly obtain a resolution of the partnership authorizing that particular partner to consummate the sale and sign the necessary documents.

Likewise, the purchasing entity should be identified before the bill of sale is completed and before the purchaser's name is filled in on the form, because if one individual or entity is shown as the purchaser on the bill of sale and another entity wishes to register the aircraft as the owner, a second bill of sale will be required to complete the links in the chain of title. It is not possible for a person or entity to register the aircraft as his (or its) if that person or entity is a stranger to the chain of owners as shown by the successive bills of sale in the airplane's history.

Blank bill of sale forms can be obtained from any FAA flight standards district office. Aircraft brokers and dealers generally have a supply on hand, and many fixed-based operators will keep a few to provide to their customers as a service. But prior to consummating the purchase of an aircraft, a buyer obviously needs to obtain the bill of sale from one of those sources.

As noted on the bill of sale form, notarization or other formal attestation of signatures is not required by federal law for recording the document with the Aircraft Registry. However, one needs to be aware of the particular requirements of the law of the state in which the transaction is occurring, to make sure that the form is completed with the formalities required by that state's law.

The title to aircraft is determined by a hybrid of both federal and state law. Generally, federal law controls, but there are nuances and subtleties where state law fills gaps or addresses questions not covered by federal law. Therefore, it is imperative that the transactional documents meet both the requirements of the FAA and the law of the state in which the transaction is occurring. Without a properly completed bill of sale, no other documents that are momentarily explained can be recorded at the Aircraft Registry, again because the buyer shown on those other documents would be a stranger to the chain of title and would create a situation wherein there would be a missing link that could not be identified by the next person searching the records.

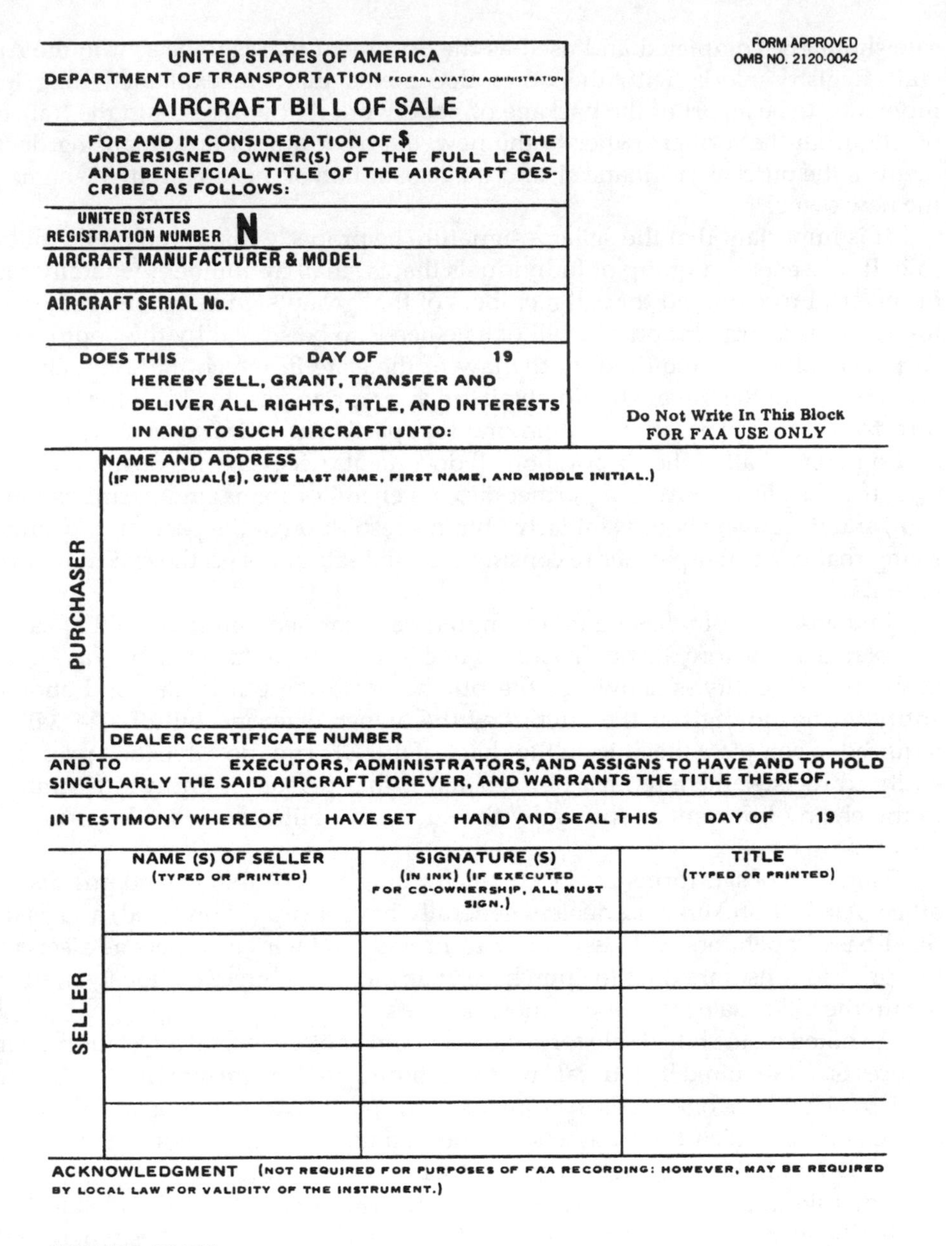

FORM APPROVED
OMB NO. 2120-0042

UNITED STATES OF AMERICA
DEPARTMENT OF TRANSPORTATION FEDERAL AVIATION ADMINISTRATION

AIRCRAFT BILL OF SALE

FOR AND IN CONSIDERATION OF $ THE UNDERSIGNED OWNER(S) OF THE FULL LEGAL AND BENEFICIAL TITLE OF THE AIRCRAFT DESCRIBED AS FOLLOWS:

UNITED STATES REGISTRATION NUMBER N

AIRCRAFT MANUFACTURER & MODEL

AIRCRAFT SERIAL No.

DOES THIS DAY OF 19
HEREBY SELL, GRANT, TRANSFER AND DELIVER ALL RIGHTS, TITLE, AND INTERESTS IN AND TO SUCH AIRCRAFT UNTO:

Do Not Write In This Block
FOR FAA USE ONLY

PURCHASER

NAME AND ADDRESS
(IF INDIVIDUAL(S), GIVE LAST NAME, FIRST NAME, AND MIDDLE INITIAL.)

DEALER CERTIFICATE NUMBER

AND TO EXECUTORS, ADMINISTRATORS, AND ASSIGNS TO HAVE AND TO HOLD SINGULARLY THE SAID AIRCRAFT FOREVER, AND WARRANTS THE TITLE THEREOF.

IN TESTIMONY WHEREOF HAVE SET HAND AND SEAL THIS DAY OF 19

SELLER

NAME (S) OF SELLER (TYPED OR PRINTED)	SIGNATURE (S) (IN INK) (IF EXECUTED FOR CO-OWNERSHIP, ALL MUST SIGN.)	TITLE (TYPED OR PRINTED)

ACKNOWLEDGMENT (NOT REQUIRED FOR PURPOSES OF FAA RECORDING: HOWEVER, MAY BE REQUIRED BY LOCAL LAW FOR VALIDITY OF THE INSTRUMENT.)

ORIGINAL: TO FAA

AC FORM 8050-2 (8-85) (0052-00-629-0002)

Fig. 3-1. Official FAA bill of sale form.

Security Agreements

When a lender provides the funds to purchase an aircraft, in all likelihood it will require that a document known as a *security agreement* be executed by the buyer/borrower. Security agreements were formerly referred to as chattel mortgages. This document gives the lender an actual property interest in the property being financed, and creates obligations of the borrower to not only pay the loan but generally to maintain the equipment, insure the equipment, notify the lender of any damage or loss to the property, and in other respects keep the equipment in an operable condition during the time period in which the loan is not yet fully repaid.

Again, for those familiar with real estate transactions, a security agreement is very much like a real estate mortgage. Also, security agreements generally give the lender the ability to take possession of the property that is serving as the security for the loan (in this case, the aircraft) in the event that the loan payments become delinquent to a specified degree. The financing of an aircraft purchase consists of two documents: *promissory note* and *security agreement*.

The promissory note is the actual promise to repay the loan to the lender. Sometimes the note and security agreement are contained within the same document, but more often are split into two separate papers. There is no requirement that the note be recorded at the Aircraft Registry, or in any other manner become public record. But the security agreement, because it gives the lender specific property rights in the aircraft, must be filed with the Aircraft Registry and become a part of the record affecting the title to each airplane.

When each buyer's loan is repaid, the lender is under an obligation to execute a release of the security agreement. Because security agreements vary greatly in their provisions, one cannot make a general statement that lenders are required to file their releases with the Aircraft Registry; and in fact, generally security agreements require only that the lender provide the release to the borrower, who in this instance is the aircraft owner. Unless the owner then sends the release to the Aircraft Registry for recordation, it will not get recorded.

Therefore, a common problem that often arises and is frequently shown in title search reports is a situation wherein old security agreements will show up as still being effective encumbrances against the aircraft. I have seen situations where decades-old security agreements are still in the chain of title and not shown as released. Usually, a contact with the lender will generate another release so that the new buyer can then send the release to the Aircraft Registry to clean up what otherwise is a serious and significant cloud on the title. But one can never assume that an unreleased security agreement that pops up on the title search is actually released. This is just one example of why a title search report needs to be reviewed by competent counsel. The time to remedy a problem such as this is when it first arises, and before the sale is completed.

With the rash of bank mergers, lending institutions going out of business, lending institutions being taken over by regulatory agents, and the like, it is a sad state of affairs for a security agreement to still be on the record when it need not be, particularly when such a security agreement shows a lien in favor of a lending institution that might no longer exist, or that might have merged years ago into another; these facts make it extremely difficult, if not nigh impossible, to secure a release.

If a security agreement exists and a release cannot be found in the records, or a new release cannot be procured from the lender, each subsequent buyer of the aircraft is taking a risk that the security agreement is in fact still valid and the underlying loan obligation was never repaid by a particular borrower. Such a chain of events could possibly result in the unexpected repossession of an aircraft because of the failure to repay a loan taken out by the owner several steps prior in the chain of title.

Again, common sense and diligence will solve many problems. Ignorance and dilatory action or total nonaction often create problems that, if allowed to fester long enough, have the potential to become insurmountable.

If you are buying an aircraft from a person who borrowed money to acquire it, your title search report will show if a security agreement is on file. It is absolutely imperative, prior to paying the purchase price and closing the transaction, that the new buyer obtain a proper and satisfactory release of the seller's security agreement, so that that release can be sent to the Aircraft Registry along with a bill of sale and perhaps a new security agreement, if the buyer is in turn also financing the purchase.

With the average age of the general aviation fleet of aircraft constantly increasing—since the supply of new aircraft is only a small shadow of its former self—these problems will only become more common and, unfortunately, more difficult to resolve. A word to the wise ought to be sufficient: *Avoid the purchase of an aircraft when the title report shows unreleased prior security agreements and releases cannot subsequently be obtained prior to the closing of the transaction and payment of the full purchase price to the seller*. As a course of last resort, if an old security agreement shows up on the title report and it becomes impossible to obtain a proper release of it, the new buyer could require an indemnity agreement from the seller, indemnifying the buyer against any loss or expense as a result of that security agreement. But this is not a total solution, and it carries risks itself.

The first of these risks is that if the old security agreement is in fact still indicative of an unpaid loan, and the lender does come forward, such an indemnity is only as good as the financial solvency of the person who gave it. A promise to indemnify a buyer against future loss and expense is worthless if the seller has disappeared, become insolvent, or gone bankrupt, or simply does not have the funds to make good on the promise of indemnity.

Another problem is that the solution is only temporary. When the new buyer later decides to sell the aircraft, the old security agreement will still show up on the title report, and then will be his problem to handle in dealing with the next buyer. Perhaps the next buyer will accept an indemnity and perhaps she will not. In any event, the presence of any unreleased security agreements needs to be carefully weighed, and the consequences determined and considered. In the last analysis, such a state of title should and almost invariably will affect the present and future market value of the airplane.

Application for Certificate of Registration

The Federal Aviation Regulations require that one of the documents that must be on board each aircraft is the *certificate of registration*. It is this paper that, when posted in plain view as required by the FAR, displays to all who have reason to come on board the aircraft the identity of the owner. Therefore, the application for a certificate of registration is the final document that must be completed and sent to the Aircraft Registry, along with the bill of sale and any releases of security agreements, or new security agreements, if the transaction involves financing. A sample application for certificate of registration appears as Figure 3-2.

Again, it is imperative that the name and identity of the purchaser under the bill of sale be precisely the same as the name and identity of the applicant for the certificate of registration. The pink copy of the application serves as a temporary registration certificate during the time that it takes the Aircraft Registry to process the recordation of the documents and issue a new certificate of registration in the name of the new owner.

Should a purchaser fail to apply for a certificate of registration, he will be operating his aircraft contrary to the mandate of the FAR, and should such violation be discovered, the owner will undoubtedly become subject to the FAA regulatory enforcement procedures fully described in Chapter 6.

When buying an aircraft through a broker, or from a dealer, the buyer can quite naturally assume that the broker or dealer will properly see to it that all the required documents are completed and sent to the Aircraft Registry. Virtually all reputable brokers and dealers do attend to this task in commendable fashion. But it is the final responsibility of the buyer to see that the appropriate paperwork, as fits the specifics of his particular transaction in purchasing the aircraft, is filed with the Aircraft Registry. Just as the pilot in command is the final authority for the conduct of a flight, the owner (new buyer) of the airplane is the person who is, in the last analysis, held responsible to make sure that his purchase is properly documented, that it is properly recorded with the Aircraft Registry, and that a new certificate of registration is both applied for and received timely, before he operates the aircraft.

CO-OWNERSHIP OF AN AIRCRAFT

Of the many reasons behind the decision to own an aircraft in conjunction with one or more other people, the one that comes to the fore most often, is simply the expense of aircraft ownership. As anyone who has ever owned an airplane will be more than happy to attest, the initial acquisition cost is just that—the initial cost. Unlike an automobile, an aircraft has continuing costs of maintaining, insuring, storing, operating, and building reserves for overhauls that can quickly outpace the amount of money that was spent in the first instance to purchase the airplane.

While it is not the purpose of this work to delve deeply into the financial aspects of aircraft ownership, one must at least give that subject preliminary treatment in order to understand why so many aircraft are owned by groups of two or more people.

FORM APPROVED
OMB NO. 2120-0029
EXP. DATE 10/31/84

UNITED STATES OF AMERICA DEPARTMENT OF TRANSPORTATION
FEDERAL AVIATION ADMINISTRATION-MIKE MONRONEY AERONAUTICAL CENTER
AIRCRAFT REGISTRATION APPLICATION

CERT. ISSUE DATE

UNITED STATES REGISTRATION NUMBER **N**

AIRCRAFT MANUFACTURER & MODEL

AIRCRAFT SERIAL No.

FOR FAA USE ONLY

TYPE OF REGISTRATION (Check one box)

☐ 1. Individual ☐ 2. Partnership ☐ 3. Corporation ☐ 4. Co-owner ☐ 5. Gov't. ☐ 8. Foreign-owned Corporation

NAME OF APPLICANT (Person(s) shown on evidence of ownership. If individual, give last name, first name, and middle initial.)

TELEPHONE NUMBER: ( ) –

ADDRESS (Permanent mailing address for first applicant listed.)

Number and street:

Rural Route: P.O. Box:

CITY	STATE	ZIP CODE

☐ **CHECK HERE IF YOU ARE ONLY REPORTING A CHANGE OF ADDRESS**

ATTENTION! Read the following statement before signing this application.

A false or dishonest answer to any question in this application may be grounds for punishment by fine and / or imprisonment (U.S. Code, Title 18, Sec. 1001).

CERTIFICATION

I/WE CERTIFY:

(1) That the above aircraft is owned by the undersigned applicant, who is a citizen (including corporations) of the United States.

(For voting trust, give name of trustee: __________), or:

CHECK ONE AS APPROPRIATE:

a. ☐ A resident alien, with alien registration (Form 1-151 or Form 1-551) No. __________

b. ☐ A foreign-owned corporation organized and doing business under the laws of (state or possession) __________, and said aircraft is based and primarily used in the United States. Records of flight hours are available for inspection at __________

(2) That the aircraft is not registered under the laws of any foreign country; and
(3) That legal evidence of ownership is attached or has been filed with the Federal Aviation Administration.

NOTE: If executed for co-ownership all applicants must sign. Use reverse side if necessary.

TYPE OR PRINT NAME BELOW SIGNATURE

EACH PART OF THIS APPLICATION MUST BE SIGNED IN INK.	SIGNATURE	TITLE	DATE
	SIGNATURE	TITLE	DATE
	SIGNATURE	TITLE	DATE

NOTE: Pending receipt of the Certificate of Aircraft Registration, the aircraft may be operated for a period not in excess of 90 days, during which time the PINK copy of this application must be carried in the aircraft.

AC FORM 8050-1 (1-83) (0052-00-628-9005)

Fig. 3-2. Sample application for certificate of registration.

Depending upon the age, condition, and time remaining before overhaul of the major power plant components, the analysis of the cost of an aircraft will vary somewhat, but not to a major degree. First of all, regardless of the age and condition of the aircraft, within reason, it will cost the same to insure as a like aircraft flown by a like pilot with similar levels of coverage. Chapter 5 is an examination of the various types of aviation insurance that are available, with definite recommendations as to which ought to be purchased by aircraft owners, as well as renter pilots.

But for the purposes to be met here, the prospective aircraft owner needs to get connected with an experienced aviation insurance broker, or deal with one of the aviation insurance companies that sell directly to aircraft owners and pilots. Companies that deal directly with their consuming public are known as direct-writing companies, whereas the majority of aviation insurers deal through independent insurance brokers.

Regardless of which way one goes, it is certainly wise to get quotes and figures from both. When the decision is made as to the types and limits of coverage to be purchased, a quotation can be obtained, and that number can be plugged into the formula for determining the cost of owning the new aerial steed.

Another number that is easy to determine is the cost of storing an airplane. The first question that demands an answer is whether the aircraft will be hangared or tied out on an open tiedown. There is not much argument that the aircraft will deteriorate at a slower rate when hangared. Indeed, in certain areas of the country known for their adverse climates, hangaring during the winter season will greatly enhance the utility of the airplane, because it will not be snow- and ice-covered after precipitation, as would be an airplane that sits out all the time. Depending upon the amount of money invested in the airplane, a person often has to take a hard, practical look at the cost of hangaring.

With hangars renting at several hundreds of dollars per month in major metropolitan areas, the cost of hangaring lower-priced aircraft quickly becomes prohibitive when put to the acid test. If you are going to purchase an airplane for only $15,000 to $20,000, and it is going to cost $300 per month to hangar it, it can quickly be seen that, at $3600 per year, approximately every 5 or 6 years you can junk the old airplane and buy a new one of similar cost and category if the hanger rent is eliminated from the budget.

However, when you consider the medium and more expensive ranges of airplanes, hangaring becomes not only an aesthetic improvement but a definite investment in maintaining and protecting the value of the airplane. Regardless of the decision made, the annual cost of storage can be quickly calculated.

Similarly, the cost of obtaining the funds to purchase the airplane is an expense that is readily calculable. If the buyer is fortunate enough to be able to pay cash, she needs only to look at the lost interest that the money would generate if safely invested, and then subtract from the annual amount of any increase that could be expected in the value of the airplane, or as an addition to the lost interest, any real depreciation (as opposed to tax depreciation) in the aircraft. When these gymnastics are complete, the cost of owning the airplane is known without yet figuring any of the operating costs or maintenance or overhaul reserve expenses.

Probably the most difficult cost estimate to generate is that for maintenance and overhaul expense. As to maintenance, a neophyte buyer needs to get good advice from a mechanic or maintenance shop in which he has trust and confidence. There are many aircraft in similar price ranges that vary greatly in the amount of normal and routine maintenance that they require.

As a rule of thumb, the fixed landing gear and fixed-pitch propeller aircraft are the simplest to maintain, and therefore the least expensive to own and to insure. As the aircraft owner moves up the ladder of performance into controllable or constant-speed propeller airplanes, and then into retractable-gear and finally multiengine airplanes, the cost of routine maintenance escalates dramatically. There are no classic formulas that will determine a budget amount for routine maintenance. Again, the best source of this information is a mechanic or maintenance facility.

Next, you must reasonably forecast the funds necessary to set aside, or plan on spending in the future, for those items that require periodic overhaul. For most airplanes, two notable items are the engine and the propeller. Merely because an engine is rated at 2000 hours between overhauls, never assume that it will reach this point. A manufacturer's suggested time between overhauls (TBO) is made with very tight parameters assumed. Such TBO generally assumes that the aircraft will fly at least 300 hours per year, and will be further subjected to scrutinous routine maintenance.

Very few things hurt an airplane more than years of idleness. So if you are looking at a particular aircraft that has had only 500 or 600 hours on its engine since the previous overhaul, and that engine has a manufacturer TBO of 2000 hours, be wary if the last overhaul was done 10 years ago. One of the most destructive factors affecting engine life is corrosion that results from lack of use. Just like exposed steel corrodes, so do the "innards" of an engine, particularly when the aircraft is not routinely flown.

Again, a thorough prepurchase inspection makes the most sense for any aircraft purchase. The inspecting mechanic can generally, after an inspection, look at the logs of the aircraft and give a reasonable expectation of the condition of the power plant and propeller, and whether it can reasonably be expected for the aircraft to ever go to the full estimate of TBO before requiring major overhaul work.

Using this information, and the number of hours remaining until the manufacturer's suggested overhaul, you must then approximate what such an overhaul will cost. Many volumes have been written simply about this subject alone. But for the purposes of these materials, it can easily be said that there are many levels of overhaul work and, as with virtually anything else, you usually get what you pay for.

If the last overhaul was extremely thorough and performed by a reputable facility, the engine can be expected to run better and longer than if that most recent overhaul was simply a "disassemble, inspect, and reassemble to service limits" type of overhaul. All these factors considered, the owner must take a stab at setting aside a certain number of dollars per flown hour to overhaul the engine and propeller when the need arises.

Next, as a new buyer you should look at the avionics equipment aboard the airplane and determine if any of it needs immediate upgrading or extensive overhaul work. Avionics also require routine maintenance, but in less sophisticated installations, I have seen many airplanes go for years between trips to the radio shop. I have also seen airplanes that cannot stay out of the radio shop more than a few weeks at a time. Again, as part of a good and thorough prepurchase inspection, all avionics equipment should be thoroughly examined, and decisions made concerning the need for immediate expense in that area.

Lastly, but certainly no less important, is the subject of cosmetics. Does the aircraft need paint? Does it need a new interior? The expenses involved in these areas are not slight. Although they can usually be postponed to a time that is convenient and affordable, the expenses cannot be delayed forever. An aircraft depends upon its paint for exterior corrosion protection, and once the painted surface becomes so bad that bare metal begins to show, obviously that corrosion protection is gone, and it is time for a visit to a reputable paint shop.

Estimates in this area are relatively easy to gather, and there is not much to cause hidden surprises. If the aircraft you are considering is in need of major refurbishment, get those estimates as part of the prepurchase inspection.

Once all these costs are calculated, the last thing to do is simply to sit down and see what it is going to cost to fly this airplane in terms of fuel and oil consumed per hour. Use the manufacturer's handbook specifications for fuel consumption, and then give it your own guess as to what fuel is going to cost, not only now, but in the foreseeable future. For those of us who remember 35-cents-per-gallon aviation fuel, the numbers that we first used to budget for our airplanes in the 1960s are obviously no longer valid and, in fact, appear humorous in retrospect.

Following this crystal balling, most people will come to the quick realization that the airplane will be much more affordable—and therefore probably get more use—if the expense of ownership can be spread among two or more individuals. While there is no fast and firm rule about how to divide the expenses among co-owners, there are basically two forms of co-ownership that accomplish some sort of expense sharing: informal co-ownership arrangements among individuals and formally organized flying clubs.

Informal Co-ownership Among Individuals

The simplest and easiest way to co-own an airplane is for two or more individuals to go together as the purchasers when buying the aircraft, and list their several names on the bill of sale, the application for registration certificate, and then the registration certificate itself.

Every co-ownership arrangement needs a written agreement covering the relationships among the co-owners, scheduling of the airplane, maintenance and improvement of the airplane, liability for damage to it, insurance coverages, sharing of expenses, and the provisions governing the buyout of a co-owner who wishes to leave the group and sell her interest in the airplane.

This written agreement is an absolute necessity. Unfortunately, during the initial excitement of forming the group, looking for an aircraft to buy, and buying it, and then the first few weeks or months of flying it, everything seems to work smoothly for most groups because emotion, excitement, and the newness of the experience control the actions of the co-owners. But eventually, when things settle down, disputes will undoubtedly arise. Knowing the resolution of those disputes in advance—by means of a properly and thoroughly prepared written agreement covering the co-ownership of the airplane—will smooth the bumps that otherwise might become moderate-to-severe turbulence during the period of time that the group owns the airplane.

Probably the first and, therefore, most important thing that must be covered is how expenses are to be shared. There are basically two competing cost-sharing theories. The first is that all the fixed expenses of owning the airplane, such as insurance, storage, overhaul reserves, and annual inspections are simply divided equally among the number of co-owners, and each person pays a percentage share. This is the arrangement that, in former years, was the most common, and is still quite often used. It works well and is fair when each of the co-owners uses the airplane approximately the same amount each year. But when there is a wide disparity in use, the basic fairness of equally sharing these expenses, which in recent years have become a larger percentage of the total cost of aircraft ownership, breaks down.

Then, it is necessary to go into a more complicated formula of cost sharing, but a formula that can accommodate the variances in flying hours put on the aircraft by the various owners. This second cost-sharing method involves some estimation during the first year of operation, and then historical data can be used to fine-tune the process in succeeding years. The method is one of taking all the costs of owning the airplane, including the fixed costs of insurance, storage, annual inspection, and overhaul reserves, and coupling them with the direct hourly expenses of fuel and oil and routine maintenance, and arriving at a grand total of cost per hour to operate the airplane.

Naturally, the group must, with some degree of accuracy, determine the projected number of hours per year that the airplane will fly, and the percentage of those hours to be flown by each person. Then, once a grand total per hour is calculated, each member of the group simply pays that amount into a common treasury for each hour of flying the airplane.

The second method tends to smooth out the inequities that result when one member of, say, a four-person group flies 200 hours per year, another flies 100 hours per year, and the remaining two fly 50 hours each. Under such an arrangement, and with such a disparity of use, the simpler method of equally sharing the large expenses connected with insurance, storage, overhaul reserve, and annual inspection is just not fair. The person or persons who fly less are subsidizing those who fly more. By calculating a grand total per hour, each pilot pays his fair share for the use to which he puts the airplane.

Either method is usable. Each particular group can decide among itself which best fits its situation.

Next, the written agreement needs to cover when the aircraft will be upgraded and what percentage of vote of the group is necessary to make major improvements such as installation of modern avionics, painting, and interior refurbishment. As with everything else involved in group activities, a certain amount of individual freedom and independence must be lost.

Generally, I suggest a two-thirds or three-quarters vote in order to spend large amounts of money on a discretionary basis, with the provision that the decision to make such improvements will be made several months before the actual expenditure would be incurred, in order that no member of the group who is of lesser financial means than others need be put to an immediate burden of coming up with what might be several thousands of dollars for such expenditures.

The written agreement also needs to address what can be a touchy subject—scheduling. There are as many ways to cover scheduling of airplanes as there are groups who co-own them. Some groups simply divide the calendar into equal periods of weeks or portions of weeks, and assign each period to an owner, during which period the aircraft is exclusively available for that particular owner's use.

I have seen groups who have no formal scheduling priorities, simply an arrangement that anyone who wants to fly the airplane calls the other owner or owners to see if they are going to use it for that particular period of time. This informal arrangement works well with very small groups of two or three owners, all of whom are easily reachable by telephone. But for busy people, or those who travel and are often out of contact for considerable periods of time, a more structured system seems to solve that problem.

There are groups who work on a system of semiexclusivity. This system allocates certain periods of time to various members of the co-ownership group with the proviso that if a pilot who is not exclusively scheduled to have use of the airplane at a particular time wants to fly it, she can contact the person who is exclusively scheduled and determine if there is a flight to be made during that day or time. If no conflict exists, then it passes to the first pilot to request that the exclusive user give up his priority for the time that the particular flight will require.

Regardless of what method suits the needs of any particular group, the arrangement needs to be covered, and covered in writing. Inevitably, conflicts will arise; and who is to say whether a particular business trip for one co-owner is more important than a vacation for another? When these matters are covered in writing and the rules of the game are known to all, the result of a conflict might be at least determined in a manner that has been agreed by the group to be fair and equitable in advance of the occurrence of the conflict itself. As with most other things in life, if a conflict can be resolved prior to its happening, the resolution seems to be less traumatic and the conflict less severe.

Probably the most important part of a written co-ownership agreement deals with the provisions necessary to accommodate the departure of one of the co-owners from the group. These provisions are commonly referred to as the buy-and-sell terms of the agreement. Intelligent people go into such a proposition knowing that it will eventually terminate, and that not all the co-owners will necessarily want to dispose of the airplane, or dispose of their ownership interests in it, at the same time. Of all the co-ownership arrangements that exist without a written agreement, *the departure of one of the members from the group* probably causes more consternation, and more litigation, than any other aspect of co-ownership among individuals.

It is axiomatic that some system must be devised to value the airplane, and each member's interest in it, at the point in time when a member of the group wishes to sell. I generally suggest to clients that, for the first year or two of the arrangement, the expense of having the airplane appraised probably is not justified, and that the initial purchase price can be used as the value of the aircraft. This system breaks down if major improvements have been made in the airplane, or refurbishment or overhaul has been undertaken during this first year or two of ownership. But in the usual situation, where the airplane is simply routinely maintained and the value of it has not been increased by any dramatic improvement or decreased by any damage, it is easy and simple to just use the initial purchase price as the value of the aircraft for the first year or two.

After that period of time has elapsed, it becomes necessary to involve some kind of appraisal to arrive at the value of the airplane. Ideally, all members of the group agree upon one individual to be the appraiser. I generally write agreements to provide that, in such an event, that one appraiser shall determine the value of the airplane and that determination shall be conclusive. If the members of the group cannot unanimously agree upon one appraiser, then the situation that routinely develops is that the remaining owners together select one appraiser and the departing owner selects a second. The two appraisers then come to their determinations of value of the airplane and, should there be any difference between them, the two appraisals are simply averaged, and that average becomes the conclusively determined value of the airplane.

If there is any loan balance outstanding, naturally it needs to be subtracted from the value so determined in order to arrive at a net figure that represents the equity of the ownership group. Then, the written agreement must deal with the method and timing of the payment of the departing owner's percentage of equity. Again, if the departure of a member occurs early in the ownership of the airplane, and particularly if there is a loan outstanding and therefore less equity, the departing owner's percentage share of the equity can be paid out almost immediately, or at least over a relatively short period of time.

But in the event that the group has stayed together a significant period of time, and if either the airplane is debt-free or the loan balance has become quite small, the payment of the departing member's share might present quite a burden to the remaining owners if it is required to be paid in a lump sum, or in a short time. In that event, I often recommend that the lump payment not be due for 6 months to a year, with one or two interim payments being made. This method gives the remaining members of the group a period of time in which to find a replacement for the departing pilot, or they may use that time span to generate the funds from other sources. In any event, this area is one that must be given a great deal of thought and that justifies, without even considering all the other important details, the production of a written co-ownership agreement.

The co-ownership agreement should also deal with the situation that can arise when some of the members of the group wish to dispose of the airplane completely and disband the group. I usually recommend that when a majority of the members wish to sell the aircraft, the minority should not be in a position to block the sale and frustrate the will of the majority. Again, the benefits of co-ownership involve the sacrifice of some individual independence, and one or two obstinate members of the group ought not to be able to prevent the disposal of the airplane if the majority wishes to do so.

In that event, when a majority vote of the group is cast in favor of disposal of the aircraft, I normally recommend to clients that the written agreement contain a provision that is known as a power of attorney, which is irrevocable, coupled with an interest, to grant the majority the absolute right to sell the airplane and to sign a bill of sale and transfer title to it even though the assent of the minority cannot be obtained.

This is not a solution for all seasons, however. Some buyers, and their attorneys, depending upon local law, might still refuse to accept a bill of sale signed under power of attorney by fewer than all the members of a co-ownership group.

This is simply one of the many reasons that co-ownership groups which are informal in nature and organization might run into difficulty in the operation of their enterprise. For that and other reasons, flying clubs have become a very popular way to co-own aircraft.

Flying Clubs

A flying club can be as formal or as informal as one wishes it to be. On the informal side, they are not tremendously different from the informal structure just discussed, although certain differences exist that merit discussion.

Unincorporated Flying Clubs

The law of most states allows for the recognition of certain unincorporated associations, particularly those that are not involved in profit-making ventures that we normally associate with commercial and mercantile activities.

The law of many states allows unincorporated associations to own property in their own names, sue and be sued in their own names, and have many of the attributes of true corporations. However, there is always a gray area, and one can never be certain that the formation of an unincorporated association will offer the benefits, immunities, and protection of proper corporate existence.

If a flying club is organized as an unincorporated association, it will usually have some charter or other organizational document that will speak to the same subjects as the written co-ownership agreement, mentioned above, used to govern the activities of the informal group. This organizational plan will sometimes call for the establishment of officers, trustees, or other such classifications of people to have managerial authority over the operation of the group.

Flying clubs, either unincorporated or incorporated, become a much easier way to deal with co-ownership, particularly when the group intends to own and operate more than one airplane. For many reasons—not the least of which is the confusion about the degree of protection and immunities present for members of unincorporated associations—I seldom recommend them as the proper structure for a flying club. Rather, the true and proper legal corporation is not that difficult to organize or maintain, and as will shortly be seen, solves most of these problems and answers most of these questions.

A True Corporation

A business corporation is extremely easy to form in almost all states. Unlike the practice several decades ago, the formation of corporations has consistently become less difficult, less demanding, and therefore less expensive with time. Today in almost all states, a corporation can be formed in a matter of a few days and at relatively nominal expense, both for the attorney fees necessary to produce the required documents and for the filing fees mandated by the general corporation laws throughout the country.

The formation of a corporation is akin to the artificial birth of a new person, because that is precisely what a corporation is in the eyes of the law. Naturally, there are certain attributes that only individuals can possess, but as far as the law is concerned, a corporation is a person and has an existence separate and apart from those individuals who own it, who are called either the shareholders or the stockholders, depending upon the nomenclature in use in the various states.

When a corporation is formed to be a flying club, the corporation itself then owns the aircraft as well as any other assets that might come into play, and the individuals involved own shares of stock of the corporation, and do not own a direct share of those assets. It is this dichotomy of identity between the corporation and what it owns, as opposed to the individuals involved and what they own, that can, at the same time, create great benefits for those individuals and cause some problems as well.

The great benefits have to do with a shielding of liability that exists when a particular member of the group, now known as a shareholder, has an accident and subjects himself and the ownership entity to liability to third persons. This entire subject is covered in detail in Chapter 8, and that material should be read and considered by all concerned prior to arriving at a decision concerning the proper business organization to use for the co-ownership of an aircraft. Naturally, these materials are intended, not to give specific legal advice to any group of individuals or any particular transaction, but simply to advise interested persons so that they are then better equipped to discuss the matters in detail with their own attorneys.

One of the negative side effects to forming a corporation is the fact that a corporation is also a "taxpayer," and even though no profit will, in all likelihood, be earned, a federal, state, and perhaps local income tax return will probably be required to be filed each year. Therefore, the legal fees and accounting fees necessary to maintain this separate "person" and report the results of its activities to the interested governmental agencies are somewhat higher than those required for the less formal types of organizations. However, these fees are generally minimal, and assuming that the various co-owners will still want to know the financial results of their operations, the costs are generally not significantly higher than the professional fees necessary to adequately operate any of the other types of co-ownership arrangements.

One of the major benefits of corporate existence is the continued viability of the corporation, and continued succession of it, through a number of shareholders, some or even all of whom depart the group over time and are replaced by others. When a corporation is used as the vehicle to acquire and retain the ownership of an airplane, there is no question but that the legal relationship is not destroyed when owners sell out and are replaced by others. Depending upon the law of a particular state, such a guarantee cannot be made when an informal organizational setup is used. This is because all states recognize the *salability* (legally called *alienability*) of shares of corporate stock. Shares can be given, inherited, sold, and in virtually any other way transferred from one person to another.

As far as a flying club is concerned, the corporate arrangement can be both a benefit and a hindrance. Perhaps the members of the group do not want strangers coming into their midst when one shareholder decides to sell her stock to someone who is not compatible with or simply not known by the others. This downside risk can easily be covered in the corporate paperwork, again through the vehicle of a buy-sell agreement, which prevents the sale of corporate shares by a shareholder except under terms that have been agreed upon in advance and to persons agreed prior to the sale. Generally, such buy-sell agreements provide that the remaining group will have a right of first refusal to purchase the shares of one who wishes to sell or otherwise transfer his interest in the corporation.

On the other hand, this alienability of the shares can be beneficial, depending on the tax-planning needs of families of co-owners and other such good-faith and justified business reasons. Again, the services of an experienced attorney need to be engaged, and legal advice solicited and followed.

The corporate format allows for the governance of the group by a board of directors and a group of executive officers. Once the group exceeds some number that is convenient for general decision making, and again of course there is no magic number, but somewhere in the neighborhood of five to eight individuals, the ability and convenience of unanimous decision making will decline and eventually break down.

If there is a larger number of co-owners owning several aircraft, it becomes totally unwieldy to require that all decisions be made by the group as a whole. The corporate vehicle allows for a board of directors and elected officers to operate the day-to-day business of the corporation, and the corporate documents can be so structured to reserve unto the entire body certain decisions of major importance, such as addition or disposal of aircraft and major improvements. One can be as restrictive or as creative as desired, because the law in almost all states gives a great degree of freedom and flexibility to the operation of such corporations, particularly when that degree of flexibility and creativity is drafted into the organizational documents at the beginning of the corporation's existence.

In summary, I generally recommend the corporate existence for the foregoing reasons. When a group of four or more individuals chooses to own an airplane together, and particularly when they intend to own more than one airplane, the corporate format offers far more benefits than detriments. For true cost efficiency and cost sharing, many flying clubs have grown quite large—for instance, more than 100 members and more than 10 airplanes. For reasons not connected with the cost-effectiveness of the arrangement, that club simply declined in size and continues into its fourth decade of existence as a smaller, yet vital organization.

It is also possible to form a nonprofit corporation to own the aircraft, and serve as the organizational form of a flying club. However, there are major differences between a typical for-profit corporation, and a nonprofit. Most of the problems associated with nonprofits surround the issues of members of the club leaving the group, and how to compensate them.

Traditionally, nonprofit corporations were most often used for social, religious, and charitable organizations such as the Girl Scouts, churches, and fraternal lodges. The theory behind nonprofit corporations allows the members of the corporation to be protected from personal liability, have continuing existence, and be managed by a group of people of lesser number than all the members.

The traditional nonprofit organization doesn't usually have an investment required to gain membership in the corporation, unless it is a fairly nominal initiation fee. But when we are considering the group purchase of an airplane, the price to get into the group won't be nominal. Each member will probably be contributing thousands of dollars to buy her "share" of the aircraft.

The law of many states prohibits the compensation of a member who is departing from a nonprofit corporation. Some states also do not allow any equity to be distributed to the members when the nonprofit corporation is eventually dissolved.

A nonprofit may work well for a group of co-owners who have been together for a significant time, when the airplane is now free and clear of any loans, and the cash flow from the use of the airplane has been applied to pay that loan. In this situation, each member may not have put any large amount of money into the original acquisition of the aircraft, and therefore doesn't have a reasonable expectancy of getting much back when it comes time to leave.

One such entity is the glider club to which I belong. It has been in existence since 1950, and only a very few of the original members are still around. The club is financially self-sustaining, and other than relying on a $200 initiation fee, it pays all its bills from the fees charged to members for use of the gliders, together with a nominal monthly dues of $11. Since no one puts any real money into the club, no one gets any of it back when he leaves. All the assets are long since paid for, and no one is liable on any loans or other obligations.

For this club, the nonprofit structure works well. But for most organizations designed to co-own aircraft, the more common for-profit corporation is quite probably the best choice of business forms to use.

OWNER-PERFORMED MAINTENANCE

Part 43 of the Federal Aviation Regulations permits pilots to perform many routine preventive maintenance operations. Specifically, section 43.3(g) of the FAR allows the holder of a pilot certificate issued under Part 61 to perform preventive maintenance on any aircraft which is owned or operated by that pilot and which is not used under Parts 121, 127, 129, or 135. These are basically the airline and air taxi sections of the regulations, and as long as the airplane is not used under those parts, a pilot may perform the preventive maintenance that is outlined in Appendix A to Part 43.

Preventive maintenance is just that, preventive in nature, and it is defined in (c) of Appendix A of Part 43 to not involve complex assembly operations and be limited to the following procedures:

1. Removal, installation, and repair of landing gear tires
2. Replacing elastic shock absorber cords on landing gear
3. Servicing landing gear shock struts by adding oil, air, or both
4. Servicing landing gear wheel bearings, such as cleaning and greasing
5. Replacing defective safety wiring or cotter keys
6. Lubrication not requiring disassembly other than removal of nonstructural items such as cover plates, cowlings, and fairings
7. Making simple fabric patches not requiring rib stitching or the removal of structural parts or control surfaces; in the case of balloons, making small fabric repairs to envelopes (as defined in, and in accordance with, the balloon manufacturers' instructions) not requiring load tape repair or replacement
8. Replenishing hydraulic fluid in the hydraulic reservoir
9. Refinishing decorative coating of fuselage, balloon baskets, wings, tail group surfaces (excluding balanced control surfaces), fairings, cowlings, landing gear, cabin, or cockpit interior when removal or disassembly of any primary structure or operating system is not required
10. Applying preservative or protective material to components where no disassembly of any primary structure or operating system is involved and where such coating is not prohibited or is not contrary to good practices
11. Repairing upholstery and decorative furnishings of the cabin, cockpit, or balloon interior when the repairing does not require disassembly of any primary structure or operating system or interfere with an operating system or affect primary structure of the aircraft
12. Making small, simple repairs to fairings, nonstructural cover plates, cowlings, and small patches and reinforcements not changing the contour so as to interfere with proper airflow
13. Replacing side windows where that work does not interfere with the structure or any operating system such as controls or electrical equipment
14. Replacing safety belts
15. Replacing seats or seat parts with replacement parts approved for the aircraft, not involving disassembly of any primary structure or operating system
16. Troubleshooting and repairing broken circuits in landing-light wiring circuits

17. Replacing bulbs, reflectors, and lenses of position and landing lights
18. Replacing wheels and skis where no weight and balance computation is involved
19. Replacing any cowling not requiring removal of the propeller or disconnection of flight controls
20. Replacing or cleaning spark plugs and setting spark plug gap clearance
21. Replacing any hose connection except hydraulic connections
22. Replacing prefabricated fuel lines
23. Cleaning or replacing fuel and oil strainers or filter elements
24. Replacing and servicing batteries
25. Removing and installing glider wings and tail surfaces that are specifically designed for quick removal and installation and when such removal and installation can be accomplished by the pilot
26. Cleaning of balloon burner pilot, and main nozzles in accordance with the balloon manufacturer's instructions
27. Replacement or adjustment of nonstructural standard fasteners incidental to operations
28. Removing and installing balloon baskets and burners that are specifically designed for quick removal and when such removal and installation can be accomplished by the pilot, provided that baskets are not interchanged except as indicated in the type certificate data sheet for that balloon
29. Installation of antimisfueling devices to reduce the diameter of fuel tank filler openings, provided the specific device has been made a part of the aircraft type certificate data by the aircraft manufacturer, the aircraft manufacturer has provided FAA-approved instructions for installation of the specific device, and installation does not involve the disassembly of the existing tank filler opening
30. Removing, checking, and replacing magnetic chip detectors

Other than these items of preventive maintenance, which, granted, are quite extensive, the maintenance work on an aircraft must be performed either by or under the supervision of a licensed mechanic, except for homebuilt and certain other experimental aircraft.

While many pilots—including your author—do not fashion themselves as even backyard mechanics, the ability to perform many of the preventive maintenance functions simply makes it legal for pilots to do cosmetic work on their airplanes that they might otherwise shy away from, or suffer the expense of having performed by certificated mechanics. Naturally, it is implicit that any preventive maintenance undertaken by a pilot (who must be at least a private pilot or higher) must be done in a good and workmanlike manner.

Appendix A to Part 43 is not a license to experiment and try to perform maintenance functions for which the particular pilot is not skilled, experienced, or otherwise competent. There have been accidents caused by improperly performed preventive maintenance, and again this whole issue of liability is fully discussed in Chapters 6 and 7. But leave it at this point simply to say that if a pilot undertakes preventive maintenance and performs the work improperly, he may well bear the brunt of the liability, should his incompetence cause an accident.

Done well and with skill, pilot-performed preventive maintenance has reduced the cost of ownership for many pilots. For those who are truly interested and mechanically inclined, many maintenance facilities and mechanics will permit the owner to assist during annual inspections, 100-hour inspections, and other inspection and maintenance functions that require the supervision of a licensed mechanic.

During the time that I was a young private and commercial pilot, working at a fixed-based operation as a flight instructor, lineman, and you-name-it, I had the opportunity to spend the best couple of years I have ever spent learning many of the aspects of our aviation industry. Even though, as previously stated, I do not fashion myself a mechanic, there is no better training in this area than to assist in the removal and overhaul of an engine, the re-covering of a fabric airplane, the reriveting of skin on a metal airplane, and similar functions to which I was exposed. I am a firm believer that the pilot who knows her airplane best can also fly it best. Part of knowing an airplane is knowing how it works, so that the pilot is in a better position to determine when it is not working properly and refer it to competent and licensed personnel.

Preventive maintenance is a small part of maintenance, but it is certainly and definitely an important part of the total process of keeping an airplane in airworthy condition. I heartily recommend that pilots frequently review Appendix A of Part 43, and avail themselves of whatever time they might have to perform those items of preventive maintenance with which each feels comfortable, not only for the benefit of having an airplane better maintained, but to become more intimately familiar with the "innards" of the machine, and as a small part of the continuing education process toward ever improving their aeronautical skills and knowledge.

We'll see in Chapter 4 that the rules concerning owner-performed maintenance are much more liberal for the builder of a homebuilt aircraft.

4

Homebuilt Aircraft

HOMEBUILT AIRCRAFT ARE NOT NEW TO THE GENERAL AVIATION INdustry. In fact, the first powered aircraft to be successfully flown in the United States was built by two brothers in their bicycle shop in Dayton, Ohio. From then, until the federal government started regulating and certificating aircraft, every airplane built could be considered a homebuilt.

Throughout the history of aviation there have been individuals who possessed the desire and ability to experiment, and create aircraft which are amateur-built. The Experimental Aircraft Association (EAA) was founded in the 1950s as an advocacy group for the rights of amateur builders of aircraft of all descriptions, and as a clearinghouse for information and assistance to builders of amateur-built aircraft. EAA is headquartered in Oshkosh, Wisconsin, where its annual convention is one of the premier, if not the utmost in, airshows and aviation events in the world. The first step for any prospective homebuilder is to join EAA.

The 1980s and thus far into the 1990s have been a time of explosive growth in homebuilt aircraft. There are several reasons, none of which is determinative or all that precisely developed for the boom in homebuilt airplanes. The cost of new factory-built airplanes is staggering. As this edition is being written, Cessna is preparing to once again build single-engine, piston-powered airplanes. Although prices haven't yet been determined, it seems that the conventional wisdom developing says that a new Cessna 172, to be built in 1997, will probably cost at least $125,000. New offerings from Piper are in manufacture now, and the simplest PA-28 is in that same general price range. Even the new two-place designs that are built by Katana and American Champion have prices that can push well past $70,000 with average avionics equipment installed.

A homebuilt airplane can cost that much too, but it usually doesn't. Considering that fact that it takes the average homebuilder several years to complete a project, the cost is spread over that time. Various parts are bought as needed, with the two most expensive components, the engine and the radio stack, generally installed last. So a homebuilder spreads out the cost of the airplane over several years, often without financing the entire price and paying large amounts of money in interest to a lending institution. In this manner, a homebuilder gets a new airplane for thousands, if not tens of thousands, of dollars less than would be spent for a factory airplane, even if the purchase price of the two are the same. Some homebuilt designs are considerably less expensive than a factory-built airplane could ever be, since many of the simpler homebuilts still use tube-and-fabric construction and may not have electrical systems and other accessories as complicated as are those installed in the products of today's airplane factories.

Since the depression in the general aviation industry began in the early 1980s, which developed into Cessna's building its last piston-powered single in 1986 and staying out of that segment of the market for 11 years, and which reduced Piper to a shadow of its former self, there hasn't been much to choose from if a pilot wanted to own a new single-engine, piston-powered airplane. Beechcraft continues to build a few Bonanzas each year, and Mooney builds its high-performance airplanes. But for the person who wants something simpler and far less expensive than the upper end of the single-engine fleet, the only offerings from factories have been coming out of specialty aircraft manufacturers like Maule, American Champion, and, off and on, Taylorcraft. Hence, building your own airplane has become the only solution to the problem for many pilots.

Homebuilding has seen a dramatic shift in what can be built, and the performance available from these airplanes. Until the 1970s, most homebuilts were either aerobatic biplanes or very simple tube-and-fabric or wood aircraft that had limited utility, compared with factory airplanes. Then came the composite revolution. As the use of composite materials gained acceptance in aircraft manufacturing, the types of airplanes that could be built in a garage, and their inherent performance, grew by leaps and bounds. Today we are at the place in the development of composite materials that a homebuilder can build an airplane that outperforms the factory airplanes, usually at lower cost and with much smaller engines.

When composites came along, the designs lent themselves to being offered in kit form by the designers who then went into business making kits. Kit building is far simpler than building from a set of plans and raw stock material. Also, a kit usually goes together much faster, producing a finished product in several years' less time than it would take to build a comparable airplane from only the plans. Since kits have come along in large numbers of different designs, there are that many more options open to those who wish to build their own airplanes.

With the increases in the numbers of homebuilts in the general aviation fleet, the FAA began to revise some of its thinking about interpretations of the regulations that govern homebuilding. We'll spend some time on this issue in a few more pages; but for now, the point is that the FAA's revising some of the rules has led kit manufacturers to be able to offer what are known as quick-build options in many of the kits now on the market. A quick-build kit implies that the manufacturer does some of the assembly at the factory. What is usually done there involves the more critical components, such as building up and riveting a metal wing box spar and doing the welding of critical parts in the case of a metal airplane, or fabricating the fuselage shells of a composite design.

Quick-build kits still often take years for the homebuilder to finish, but the FAA has come to realize that certain parts of the aircraft, especially critical load-bearing parts of the airframe, can be built better at the factory, and the resulting airplane may be all that much more safe. Quick-build kits have opened the avenue of homebuilding to more pilots, particularly those who have the time, but perhaps not the utmost in mechanical skill which is really needed to build an airplane from only the plans and raw materials.

For those who have the time and ability, building an airplane can be one of the more satisfying pursuits in life. If you read EAA's magazine, *Sport Aviation,* you will see in its pages some examples of absolutely outstanding craftsmanship, and the pride on the faces of the builders of those homebuilts.

In the remainder of this chapter, we'll concentrate our discussions on the legal issues surrounding four aspects of building your own airplane. These subject areas are building and certifying an aircraft, maintaining it, operating and insuring it, and finally, the concerns that abound when a homebuilder wants to sell or otherwise dispose of the finished airplane.

BUILDING A HOMEBUILT—PLAN FIRST

There is an old adage among carpenters—measure twice, cut once. This advice will be applicable to the physical building of an airplane, but is also appropriate to follow before you cut the first piece of material, or order a kit. If the job is well planned from the beginning, and all the required steps are followed to assure yourself that the FAA will approve the aircraft, once it's built, you'll end up with an airplane. Otherwise, you'll be in a quagmire, and all that will result could well be a very expensive aggregation of parts that can never be legally flown.

The first two steps can be taken in whichever order you choose. Join EAA, and go see the FAA flight standards district office (FSDO) nearest you. The visit to the FAA office is vital, and should never be omitted, regardless of how much you think you may know about the homebuilding process.

The eventual inspection of your airplane will be given by either an FAA inspector or a designated airworthiness representative (DAR). Unfortunately, there is quite a bit of variance in how some of the homebuilding rules are interpreted and applied from one FAA region to the next, or even from one FSDO to another within the same region. If you once built an airplane while living in Texas, and you've subsequently moved to the northeastern United States, don't count on your experience in Texas being the same as it will be now.

Sit down at the FSDO with an airworthiness inspector, and explain the project that you want to undertake. Ask what kind of builder's log the inspector wants to see, once the airplane is finished. The rules require that you keep a log, noting each phase of construction by date and assembly step. Some inspectors demand a more detailed log than others; a few will want the substantiating data recorded in their own way.

The rules also dictate that you must eventually show the inspector or DAR a photo album that depicts the various stages of building the aircraft. The FAA wants to be able to ascertain that you built the airplane, and didn't just buy it from someone else who never obeyed the regulations. Inquire about the number of photos needed, what kind and size, how detailed they need to be, whether you need to be shown in them, and any other such question that you can conceive. Once the airplane is built, you can't go back and fill a void in the photo album requirement, so find out exactly what your FAA office wants to see.

While at the FAA office, pick up a copy of Advisory Circular No. 20-27D, or whatever iteration of the AC is current at the time. This AC outlines all the steps necessary to certify and operate an amateur-built aircraft. Read it thoroughly, and keep it handy. If you are a true neophyte at homebuilding, it would be wise to get a copy of the AC and go through it before you meet with the FSDO inspector, so that you can add any questions about the contents of the AC to your agenda for the meeting. There are a number of other advisory circulars listed in the appendix of AC 20-27D that are applicable to the homebuilding process, so get them too.

EAA is an organization to which almost every homebuilder belongs, as well as thousands of "airplane nuts" who have no intent of ever building their own aircraft. Everyone who is even thinking of building an aircraft should join. EAA's monthly magazine, *Sport Aviation,* is full of helpful hints and technical advice, and is worth more than the cost of joining EAA.

The organization has a complete member services department, to which homebuilders can turn with every imaginable question. If the staff at Oshkosh can't answer the question, they will tell you where to go for a resolution of your query. EAA has a program which they call Technical Counselors. These counselors are members who have built their own airplanes, and are folks on whom you can lean for a reasonable amount of advice, and sometimes a limited amount of hands-on assistance during your project. They aren't there to build your airplane, or to be constant companions throughout every step; but they volunteer their time to help fellow homebuilders through particularly tough portions of a project.

EAA has also appointed some of its members as counselors to help pilots get the proper training and proficiency to safely fly the airplane once it's built. The early flight test program, about which I'll say more soon, is no place for a marginally qualified pilot to be the first to fly any homebuilt. These counselors often serve as test pilots, or at least give the builder some hints of what to expect, and of what to be wary, during those all-important first few flights. Many pilots trained in the past two or three decades haven't even sat in a tailwheel airplane. So don't be trying to teach yourself to fly one, while at the same time flying a homebuilt for the first time, if you are among their ranks. Many great homebuilt designs and kits are tailwheel airplanes, and you'll need currency in that configuration of airplane before even considering doing your own flight test program.

After you've done your homework—joined EAA and been to visit the FSDO inspector—you can start to consider the other aspects and rules surrounding amateur-built aircraft.

THE 51 PERCENT RULE

FAR 21.191 is the section of the regulations which generally governs experimental aircraft. Every potential homebuilder needs to read it carefully, and then reread it a few more times. This relatively short FAR deals with the purposes for which experimental airworthiness certificate are issued.

Section 21.191(g) covers certification for operating amateur-built aircraft. According to this section, an experimental airworthiness certificate may be granted for "operating an aircraft the major portion of which has been fabricated and assembled by persons who undertook the construction project solely for their own education or recreation."

That is all there is to the regulation that authorizes the entire homebuilt industry and movement within the United States. It looks simple at first, but like most legal issues, it is anything but that in application. First, the person or persons who built the aircraft must fabricate and assemble the major portion of it. This dictate is interpreted by the FAA to mean that the builder must do at least 51 percent (being the major portion) of the project. But what does that mean? Also, the project must be undertaken for the education or recreation of the builder. That part isn't so susceptible of different meanings.

The idea that the builder must do the major portion of the fabrication and assembly has been interpreted in different ways at different times. The rapid increase in kits required the FAA to think more about it than was needed in the days when all homebuilts were made from plans and raw materials. For in that era, unless you were an outright cheat and liar, you couldn't do anything else than build the major portion of the aircraft yourself. But now we have kits that have major components fabricated at the factory and, unfortunately, a cast of characters who build airplanes under a compensatory arrangement for the real owner.

Labor alone doesn't count for purposes of meeting the 51 percent rule. You could easily spend 51 percent of the total time involved in a project doing very few of the fabrication and assembly operations. If only labor counted, a person could spend 51 percent of the time just doing an interior, installing an exotic avionics package, applying multicoat paint schemes, or completing a myriad of other steps that, when the airplane is viewed as a whole, didn't add up to that much real nuts-and-bolts fabrication and assembly.

So now the FAA has an interpretation of the 51 percent requirement that focuses upon the various tasks involved in building the aircraft. The builder must still do 51 percent of the labor involved in bringing the kit to fruition as an airplane, and that is seldom a practical problem, since homebuilding is such a time-intensive endeavor in the first place. In addition to the labor requirement, the person who claims to have built the airplane must be able to document that she performed at least 51 percent of the building operations or tasks.

So the builder doesn't have to drive 51 percent of all the rivets in the entire airplane. He gets credit for this task by demonstrating that he can drive wing, fuselage, and tail rivets. This allows the builder to claim credit for the riveting task. The total of all tasks required to complete the airplane must leave 51 percent of them in the amateur's hands; the kit factory can do 49 percent. The FAA is constantly evaluating kits for compliance with the 51 percent rule, and a list of approved kits can be obtained from your local FSDO. If you build one of the approved kits, and do it yourself, you'll pass this requirement.

Another problem plaguing the homebuilt aircraft industry is the presence of professional builders who, for a fee, will see any homebuilt through from dream to test flight. Since most of these people operate underground, and usually illegally, they don't apply for the airworthiness certificate themselves. Many tell the customer to lie, and sign the FAA-required affidavit which says that the customer built the aircraft. Get caught in that deception, and you're subject to reducing the national debt by a potential $10,000 fine; further, you may get to spend a 5-year vacation as a guest of the federal government. At least they've closed Alcatraz, so you won't go there.

The "hired gun" builder has cropped up in recent years for several reasons. The most obvious of these is the increasing complexity and sophistication of some modern homebuilt kits. In the mid-1990s, it is possible to spend over a quarter of a million dollars on the most exotic of the homebuilts. Aircraft like these, at least one of which is pressurized, are terribly time-intensive to really build yourself. Couple that fact with the notion that more than a few of the people who can afford them hardly have the time to spend thousands of hours in a garage or shop themselves, and may be businesspeople who've never done anything more mechanical than use a simple screwdriver, and you've got the recipe that feeds the commercial "homebuilding" operation.

The FAA and EAA are both so concerned about this problem that they've been working together for some time to come up with a solution. The EAA is properly very jealous about keeping the right for amateurs to build airplanes. No ethical person associated with the homebuilding industry wants to see crooks ruin it for everyone else. The FAA has come up with a new advisory circular outlining what help a homebuilder can get, and still meet the 51 percent rule. The FAA calls this kind of help "commercial assistance."

The new AC doesn't prohibit commercial assistance, but makes it clear that the homebuilder must still meet the 51 percent rule, after considering the tasks performed by the commercial operation. The big problem arises when the amateur gets help on a kit that the FAA has already approved as meeting the rule. Most of these kits, particularly in their quick-build versions, barely meet the rule. Any amount of help may put the homebuilder in the position of not complying, and thereby making the finished product ineligible for an amateur-built experimental airworthiness certificate. The AC also says that certain tasks, such as custom paint jobs or very complete avionics packages, are operations that may qualify for help, since they aren't "fabrication and assembly" of the aircraft itself. The builder is also allowed to get instructional assistance on specified construction tasks. If you anticipate getting any assistance whatsoever on a homebuilt, get a copy of the new AC. Then meet with your FSDO inspector, and come clean about just what it is that you intend to ask others to do. An ounce of this kind of prevention of later certification problems is worth many pounds of cure, especially when there may not be any cure after the airplane is built.

If you don't comply with the 51 percent rule, then the aircraft is eligible for certification only under the experimental-exhibition category. Certification under those rules allows the airplane to be flown from the owner's home airport, and maybe to and from approved airshows. The result is that the utility of the aircraft may be so hampered as to make it nearly worthless, both to the owner and to any subsequent purchaser.

CERTIFICATION STEPS

Let's now assume that you will comply with the 51 percent rule, and won't cheat on the construction of the aircraft. The FAA, through an FSDO inspector or a DAR, will still have to certify your airplane when it's built. Before it can be flown, it must be issued an experimental airworthiness certificate, be assigned a registration number, and be registered, just the same as a factory-built aircraft.

At some time during the construction project, but at least 3 or 4 months before you think the airplane will be ready to fly, you need to write to the FAA in Oklahoma City and obtain a reserved registration number, which will be the N number that must be displayed on the sides of the airplane, or elsewhere if you've built some kind of aircraft other than an airplane or glider. The N number must be affixed before you apply for the airworthiness certificate, as must the data plate, which identifies the aircraft, its manufacturer (you), and its type and serial number. FAR 45.13 specifies the information required on the data plate.

In days gone by, the FAA used to perform periodic inspections of homebuilt aircraft at at least one stage of construction, and potentially others. This was called the "precover inspection," and was made prior to the structural members of the aircraft being covered with either fabric or metal skin. After decades of this practice, the requirement for a precover inspection was eliminated. Now the FAA inspector or DAR needs to inspect the aircraft only once, before the first flight. In order to get your airworthiness certificate, you must present a complete aircraft, ready for flight. You'll be asked to remove fairings, cowlings, and inspection plates so that the inspector can see how well you've built the airplane, but it must be finished.

During this inspection, the builder is required to present not only the airplane, but also the builder's log, photo album, receipts, and anything else which the inspection reasonably requests. The inspector is checking two things—the physical construction of the aircraft and its safety for flight—and is ascertaining that you actually built it.

If the airplane passes the inspection, and if you have your registration certificate properly displayed, data plate and N number affixed, the next step is issuance of the airworthiness certificate and operating limitations.

The operating limitations will be in two basic parts. First, you'll be assigned a specific flight test area, and number of hours that the airplane must spend in that area before it can be flown out of it. The flight test area will be small, generally within 25 miles of your home base. If you've built an airplane which used a type-certified engine-propeller combination, the flight test time will be 25 hours. Otherwise, it'll be 40 hours. During this time, you will be tweaking many things anyhow, since few homebuilts perform perfectly on the first few flights. Engine baffling and cooling, ignition timing, and the like often need alterations during the test period. During the test time, only required crew members may occupy the aircraft, which means no passenger carrying or flight instruction.

Once the flight test time has been flown off, the operating limitations will allow you to fly the airplane much the same as you would a factory-built aircraft, except that an amateur-built aircraft can never be flown for compensation or hire. Now that you've planned, built, and obtained certification of an amateur-built aircraft, let's see how it must be maintained.

MAINTAINING HOMEBUILTS

There are many reasons for building one's own airplane, and just as many aspects of pride and pleasure derived from successfully completing the process. One of these is that the builder will generally qualify to maintain the aircraft, even if he isn't already a certificated mechanic.

The FAA will generally issue what is known as a repairman certificate to the primary builder of the aircraft. If you've built the aircraft alone, complying with the 51 percent rule, a repairman certificate is almost automatically issued. If a group of individuals does the building, the group leader may apply for the certificate, but its issuance isn't guaranteed.

The FSDO inspector who has authority to issue a repairman certificate is interested in ascertaining if the applicant has the needed skill to determine whether the aircraft is in a continuing condition for safe operation. When an airplane is built by one person alone, that skill is virtually certain to exist. But when groups of individuals build, there may not be one person who can demonstrate enough global knowledge about the entire airframe and engine to be considered by the FAA as skillful enough to be certificated to inspect and repair the airplane.

Often groups, such as an EAA local chapter, will undertake the construction of a homebuilt, and the FAA doesn't discourage such efforts. If there is a group leader who possesses the general knowledge and skill about the entire aircraft, and its construction, perhaps that individual can persuade the FSDO to issue a repairman certificate. If the job of building the aircraft was divided among people in the group by tasks, maybe no one will get the repairman certificate. If Joe did the welding, and knows nothing about fabric or engines, Susan did the fuel system installation but knows nothing about basic structures, Henry did the fabric covering but likewise is ignorant of engines, and Sarah installed the engine but doesn't know a whit about fabric or structures, there is nobody in this hypothetical group who has the requisite skill to serve as repairman.

The beauty of having a repairman certificate, which, applied to homebuilt aircraft is always limited to that specific aircraft, is that the repairman can function as mechanic and inspector for all repairs and condition inspections. The certificated repairman not only can maintain the airplane, but also is authorized to perform and certify the annual inspections.

In order to get a repairman certificate, you must apply to the FSDO after the airplane has completed its required flight hours in the designated test area. There is a particular form that the FSDO will give you to complete, which cannot be submitted by the builder until after the required flight test hours are completed. Even if you've had a repairman certificate for another homebuilt, you must go through the application and issuance process to get recertified for this particular amateur-built airplane. If you are successful in your application, then the certificate will be reissued, with an added limitation for the newly built aircraft.

If you purchase a homebuilt that has already had its airworthiness certificate issued, and was built by someone else, you won't be able to get a repairman certificate for it. In that case, it will have to be maintained and inspected by certified maintenance personnel, or by an approved repair station, just the same as if the airplane were factory-built. With all the increased activity in homebuilding, there are more used homebuilts now on the resale market than ever before. The lack of being able to get yourself certified as a repairman for an aircraft which you buy after it has been constructed and certified isn't reason to ignore the possibility of acquiring a homebuilt, but you won't enjoy the obvious cost savings inherent in being able to be your own mechanic and inspector.

OPERATING AND INSURING YOUR HOMEBUILT

There are some differences in the day-to-day operation of homebuilt aircraft as compared with how you can use a factory-built aircraft. Some are relatively insignificant; others are major. The most obvious restrictions on the flying of homebuilts, after the test period is finished, come from the operating limitations that the FAA issues with the experimental airworthiness certificate. A set of typical operating limitations is set forth in Figure 4-1. This list comes right out of AC 20-27D. Let's go through the important limitations.

Amateur-built aircraft cannot be used for hire or compensation in passenger- or cargo-carrying operations. So they are restricted to operations conducted under Part 91 of the FAR. They can't be used for charter flying under Part 135, or for flight training under Part 141, except in the rare instance that an owner would get approval to enroll herself in a Part 141 school, and use her airplane for her own training.

Homebuilts are not supposed to be operated over densely populated areas or in congested airways, except as necessary for takeoffs and landings. With the appropriate equipment installed, homebuilts can be certified for IFR flight. That creates an oxymoronic conflict in the rules. When flying IFR, at least for as long as the IFR system is based upon VOR navigation on airways, it would be nigh impossible to fly an IFR flight and at the same time avoid congested airways. This is a regulation that exists, but isn't rigidly enforced. Likewise, amateur-built aircraft do fly over densely populated areas, but the pilots of them do so at the peril of being charged with exceeding the airplane's operating limitations.

Whenever an amateur-built aircraft flies into or out of an airport with an operating control tower, the word *experimental* must precede the N number in the first call to the controller. If you build or buy one of the more popular kits like a Glasair, you don't identify yourself as a Glasair, as you would call yourself a "Cessna" if you were flying a Cessna 172. If the controller asks what type of homebuilt you happen to be flying, obviously you need to answer the question. There may be some change in this notion in the future, since experimental homebuilt aircraft now span such a wide spectrum of performance capabilities. When this restriction was first put in operating limitations for homebuilts, they were almost always simple airplanes with relatively slow speeds. Now a pressurized Lancair IV-P could

APPENDIX 9. SAMPLE LIST OF OPERATING LIMITATIONS

THESE OPERATING LIMITATIONS SHALL BE ACCESSIBLE TO THE PILOT

EXPERIMENTAL OPERATING LIMITATIONS
OPERATING AMATEUR-BUILT AIRCRAFT

REG. NO. ____________ SERIAL NO. ________
MAKE ____________ MODEL ________

Phase I. Initial Flight Test in Restricted Area:

1. No person may operate this aircraft for other than the purpose of operating amateur-built aircraft to accomplish the operation and flight test outline in the applicant's letter dated ________ in accordance with FAR section 21.193. Phase I and II amateur-built operations shall be conducted in accordance with applicable air traffic and general operating rules of FAR Part 91 and the additional limitations herein prescribed under the provisions of FAR section 91.42 (new FAR section 91.319).

2. The initial ______ hours of flight shall be conducted within the geographical area described as follows:

____
____

3. Except for takeoffs and landings, no person may operate this aircraft over densely populated areas or in congested airways.

4. This aircraft is approved for day VFR operation only.

5. Unless prohibited by design, acrobatics are permitted in the assigned flight test area. All acrobatics are to be conducted under the provisions of FAR section 91.71 (new FAR section 91.303).

6. No person may be carried in this aircraft during flight unless that person is required for the purpose of the flight.

7. The cognizant FAA office must be notified and their response received in writing prior to flying this aircraft after incorporating a major change as defined by FAR section 21.93.

8. The operator of this aircraft shall notify the control tower of the experimental nature of this aircraft when operating into or out of airports with operating control towers.

Fig. 4-1. Sample operating limitations for an amateur-built aircraft.

APPENDIX 9. SAMPLE LIST OF OPERATING LIMITATIONS (CONT'D)

9. The pilot-in-command of this aircraft must, as applicable, hold an appropriate category/class rating, have an aircraft type rating, have a flight instructor's logbook endorsement or possess a "Letter of Authorization" issued by an FAA Flight Standards Operations Inspector.

10. This aircraft does not meet the requirements of the applicable, comprehensive, and detailed airworthiness code as provided by Annex 8 to the Convention on International Civil Aviation. This aircraft may not be operated over any other country without the permission of that country.

Phase II:

Following satisfactory completion of the required number of flight hours in the flight test area, the pilot shall certify in the logbook that the aircraft has been shown to comply with FAR section 91.42(b) (new FAR section 91.319). Compliance with FAR section 91.42(b) (new FAR section 91.319) shall be recorded in the aircraft logbook with the following or similarly worded statement:

"I certify that the prescribed flight test hours have been completed and the aircraft is controllable throughout its range of speeds and throughout all maneuvers to be executed, has no hazardous operating characteristics or design features, and is safe for operation."

The Following Limitations Apply Outside of Flight Test Area:

1. Limitations 1, 3, 7, 8, 9, and 10 from Phase I are applicable.

2. This aircraft is approved for day VFR only, unless equipped for night VFR and/or IFR in accordance with FAR section 91.33 (new FAR section 91.205).

3. This aircraft shall contain the placards, markings, etc., required by FAR section 91.31 (new FAR section 91.9).

4. This aircraft is prohibited from acrobatic flight, unless such flights were satisfactorily accomplished and recorded in the aircraft logbook during the flight test period.

Fig. 4-1. (*Continued*)

APPENDIX 9. SAMPLE LIST OF OPERATING LIMITATIONS (CONT'D)

5. No person may operate this aircraft for carrying persons or property for compensation or hire.

6. The person operating this aircraft shall advise each person carried of the experimental nature of this aircraft.

7. This aircraft shall not be operated for glider towing or parachute jumping operations, unless so equipped and authorized.

8. No person shall operate this aircraft unless within the preceding 12 calendar months it has had a condition inspection performed in accordance with FAR Part 43, appendix D, and has been found to be in a condition for safe operation. In addition, this inspection shall be recorded in accordance with limitation 10 listed below.

9. The builder of this aircraft, if certificated as a repairman, FAA certified mechanic holding an Airframe and Powerplant rating and/or appropriately rated repair stations may perform condition inspections in accordance with FAR Part 43, appendix D.

10. Condition inspections shall be recorded in the aircraft maintenance records showing the following or a similarly worded statement:

"I certify that this aircraft has been inspected on (insert date) in accordance with the scope and detail of appendix D of Part 43 and found to be in a condition for safe operation."

The entry will include the aircraft total time-in-service, the name, signature, and certificate type and number of the person performing the inspection.

_________________________ _________________________

Aviation Safety Inspector Date Issued

____

Office Designation

Fig. 4-1. (*Continued*)

be in the same airport traffic pattern with a Kitfox, and each would have identified itself only as "Experimental 1234 Victor" on the first call up. If you fly a more advanced homebuilt, there is nothing wrong with beginning your radio talk with something like "Experimental Lancair Four 1234 Victor."

You have to have a pilot certificate to fly the aircraft. Nothing about operating a homebuilt relieves the pilot from the requirement to personally possess the applicable pilot license and rating to fly the airplane. For most homebuilts, a recreational or private certificate is all that is needed. Naturally, if you intend to seek IFR approval for the airplane, you'll need an instrument rating to fly it IFR. The only exception to the need to be licensed is if the end product of what you build is an ultralight, for which no pilot certificate is needed. While this book doesn't cover ultralights, there is no airworthiness certificate issued for them either. Ultralight operations are governed only by Part 103 of the FAR, which is merely a skeletal set of regulations.

You must advise each passenger whom you carry that the aircraft is amateur-built. This task is usually accomplished by installing a small placard, in plain view of the passenger seat, to that effect. This advice probably won't help you much if you have an accident, and an injured passenger later files a civil lawsuit against you; but at least you'll have complied with that part of the aircraft's operating limitations.

If you want to use the airplane for dropping parachute jumpers or towing gliders, those operations must be specifically approved, and the prohibition against such flights must be removed from the operating limitations. I doubt that that will be of concern to very many owners of homebuilts.

Lastly, the aircraft must have an annual inspection, just as does a factory-built one. The wording that is needed in the aircraft's maintenance records is a little different, but the inspection itself is not. The inspection may be preformed by either the builder, if he has a repairman certificate for that airplane, an airframe and powerplant mechanic, or an FAA-approved repair station. Annuals on homebuilts create a problem for some mechanics and repair stations.

No matter how well built the aircraft may be, there is no way that during an annual inspection a mechanic can ascertain everything about how the aircraft was assembled. No matter how many inspection plates and openings are installed, you can't see every weld and every other critical assembly point during an annual inspection. When a mechanic signs off the required entry in the aircraft's logbook or other maintenance record, that mechanic's fortune is on the line. If a subsequent accident occurs that can be alleged in any way, even remotely, to be associated with the condition of the airplane, the mechanic will likely get an invitation to the courthouse party when someone files a lawsuit over the accident.

I have advised clients who are mechanics, or who are approved repair stations, to tread with caution when asked to perform an annual inspection on a homebuilt, for that very reason. Life is full of risks, and we can't eliminate them all. However, if I were in the maintenance business, I would be careful to make sure that I knew the builder's skill and work habits before I annualed somebody else's homebuilt. If you have any desire to purchase a homebuilt from a builder, or other subsequent owner, have a good talk beforehand with the mechanic or maintenance facility where you plan to have the aircraft maintained and inspected.

You might find that your favorite mechanic may well shy away from working on it, let alone annualing it. We talked about prepurchase inspections in Chapter 3. If you are looking at a homebuilt to buy, a prepurchase inspection is even more vital. At least when you buy a Piper, Cessna, Mooney, Beech, or other factory-built airplane, you know that when it was rolled out of the factory doors, it was put together properly, and that it conformed to the approved type certificate. You've got no idea what you're getting when you buy someone's homebuilt. Another reason to do a good prebuy inspection is that it gives your mechanic an opportunity to look at the airplane, and decide whether he'll have any future reluctance to work on and inspect it.

Homebuilts are not difficult to insure these days, with a few exceptions. As these pages are written, at least one of the major insurers of homebuilts is not current agreeing to insure a few of the very high performance amateur-built airplanes. Several aviation insurance companies do insure homebuilts, and the premium charges don't differ measurably from those for factory-built airplanes of like value, with similar limits of coverage. But check with your insurance agent or broker first to see if the type that you are considering will be difficult to insure or, worse yet, not insurable at all.

Homebuilts do have a safety record not quite as stellar as factory-built airplanes. There are many causes, most of which relate to pilot error, not the airplanes themselves. However, since homebuilts don't have to comply with the certification regulations, such areas as stability and stalling speed of an amateur-built design may be of a nature that would make the airplane ineligible for a normal airworthiness certificate. Some of the high-performance homebuilts do have high wing loadings, high stalling speeds, and stability characteristics that make them more demanding of proper handling, and call for a higher level of pilot skill than a Cessna 150 requires.

Homebuilts that use odd engines, not normally used in aircraft, may encounter difficulty in getting insurance, or the premium may be higher than you would otherwise expect, or the offered limits of coverage may be low. Difficulty might also be expected if you're trying to insure an aircraft which isn't very popular in the experimental ranks, and about which there isn't much track record of safety, or lack thereof. Most insurers don't like to be pioneers—the entire concept of insurance is risk sharing, in which an underwriter can intelligently assess the risk that is requested to be insured. If an underwriter can't find out much about the risk, she may well pass on the opportunity to insure it.

Before we talk about the problems encountered when selling a homebuilt, let's look at another movement in the general aviation industry which is akin to homebuilding.

RESTORING THE OLDER AIRPLANE

There is an entire area of endeavor which has its own hordes of practitioners, and that is the restoration of older, classic aircraft. There is no precise definition for what constitutes classics, but most people think of them as airplanes built right after the end of World War II, up through the early to mid-1950s.

Civilian airplane production ground to a halt after Pearl Harbor, as the entire nation shifted to wartime production. Piper, Cessna, and Taylorcraft still built airplanes. The Piper J-3 gave up its yellow paint for olive drab, and became the Army L-4. Cessna built the famous T-50 and UC-78, the later affectionately known to those of us who have flown one as the Bamboo Bomber, since it was primarily made of wood. Taylorcraft built the L-2, an airplane very much like the L-4 from Piper.

When the war ended, there were thousands of military and naval aviators already trained, whom the factories thought would be standing in line to buy new airplanes, and keep flying. We saw a tremendous boom in light-plane production from 1946 to 1948. By 1949 production started slowing, and the heyday ended early in the 1950s. Why the boom lasted for so short a time is still the subject of conjecture, but the plain fact is that the factories were wrong in their market analysis. The former military and naval types didn't buy airplanes in droves; and a good number of them never flew again.

Maybe the association of airplanes with combat, death, and warring miseries ruined flying for some of these veterans. Other reasons could easily be that many weren't excited about flying an airplane whose maximum speed wasn't even up to the stalling speeds of front-line military and naval combat aircraft. Thousands of returning veterans had things like finishing their education, getting a job, and starting delayed families on their minds, and put aviation on the back burner. During my days as an active flight instructor in college during the 1960s, I had the pleasure of flying with several of this class of pilots, who, at the time, were rediscovering flying after a 20-year layoff since the end of the war. By then, they had their lives in order again, and some of them came back to the airport.

The result of this misjudgment by the aviation manufacturers was, and still is to a great degree, that thousands of airplanes were built during those few years in the late 1940s. Many are still around today, and can offer a great amount of enjoyment as sport aircraft, and are also constantly appreciating in value.

Since all the classic airplanes are at or near 50 years old, they need periodic restoration. Engines need to be overhauled, airframes need a thorough inspection for corrosion and often repairs, and the fabric covering on most classics requires replacement every so often. Restoring a classic has benefits and downsides not present in building an amateur-built aircraft.

First and foremost, you are dealing with a type-certified, factory-built airplane. Therefore, you cannot restore it yourself, and legally sign the paperwork to return it to service, unless you are a certificated mechanic, who also has inspection authorization. These aren't homebuilt airplanes, and they have to be maintained, repaired, and restored by licensed maintenance people.

You can do a great deal of the work without a mechanic's certificate. What is required is that a mechanic supervise your work, and that one with inspection authorization return the aircraft to service after the restoration is complete. There is no such thing as a repairman's certificate for a restored factory aircraft, so it will always have to be maintained and inspected by certified personnel.

Depending on your level of skill, and how well you know a licensed mechanic, the amateur can completely restore a classic under supervision, the depth of which will vary according to those factors. Many a restoration has, for all practical purposes, proceeded like a homebuilding job. If you know what you're doing, and the licensed mechanic trusts you, you may have very few visits to inspect your work. Without those operable conditions, you may need guidance every step of the way.

There are many benefits to restoring a classic. Even if the airplane has to be stripped to the tubing, and the engine completely overhauled, most restorations will consume far less time than building an airplane, even from a kit. Parts are readily available for almost all the classics, and all the restorer usually has to do is to buy and install them, rather than fabricating pieces from raw stock. Occasionally a part will have to be made, but not often. Unless the airplane has been damaged by an accident, most of the airframe will be reusable, after a good cleaning and inspection.

When the job is finished, you have a certified aircraft with known flying qualities, and predictable handling. Unless you botch something during the restoration, you know what you've got. Even the finest job of building a homebuilt will result in an airplane with some unknowns about it. Unfortunately, the annals of homebuilding have several pages full of tragic accidents that happened during the first flight of a homebuilt. One of the reasons for this sad statistic relates to the time that it takes to build the average homebuilt. All too frequently, a homebuilder will spend all the available time, and money, in the garage. Several years later, he'll then attempt a first flight, lacking any notion of currency or recent proficiency as a pilot. The result is predictable.

Restoring a classic is usually less costly than building a homebuilt of equivalent performance. Again, most of the airframe is normally all right. When you build an airplane from either scratch or a kit, you're buying every last nut and screw at new prices. When restoring, you buy only what is needed to put the airplane back in condition. If a person has the skill to build an airplane, restoring one should not be difficult. But if you don't know the basic skills involved, and have to hire a mechanic to do all but the most elementary jobs, your restoration project can get very expensive very quickly.

Once a classic is restored, and restored to pristine condition, it should be very easy to sell if you keep it hangared, and maintain it as well as it was restored. Like-new J-3 Cubs are pushing over $30,000 in the mid-1990s. Not bad for an airplane that could have been had in the 1970s for around $4000–$5000. Classic airplanes have been tremendous investments for the last 25 years; most used airplanes have been too, but the more modern ones haven't achieved the gains in value that the classics have.

SELLING THE HOMEBUILT

Homebuilt airplanes pose special considerations when the builder wants to sell the product of so many years' labor. The most important of these is the liability for future accidents that someone else may suffer. While a homebuilder is not a manufacturer in the sense that Cessna and Piper are, problems could be visited upon an amateur builder years after the airplane is sold.

To be strictly liable for one's products, the manufacturer has to place them in the stream of commerce, which isn't applicable to the ethical homebuilder, although it may apply to the hired-gun commercial operation discussed earlier in this chapter. What we have to keep in mind is that we live in a litigious society, and when someone is hurt or killed in an accident, a lawsuit usually follows. When a homebuilder sells an airplane, there are some protective measures that can be taken, but they are by no means foolproof, and they don't guarantee immunity from liability. If a homebuilder is sued after the airplane is sold, the suit will probably allege negligence in the building of the airplane, or the failure to warn the purchaser of any dangerous or hidden qualities of the aircraft.

It sounds like an old saw to keep saying that we cannot eliminate all risks in life, but that is true. I have heard some lawyers advise homebuilders never to sell the airplane whole, but to take it apart, and sell only the parts, and then to different people, so no one person could reassemble it. That's certainly a safe approach, and reduces the risk of selling to as little as it can be. Only you can decide what level of risk you're willing to accept.

Regardless of the measures taken to protect against a suit by the purchaser, about which we'll speak in a moment, there isn't much that can effectively be done to insulate the builder from a suit by a passenger who is hurt or killed while the airplane is in the hands of the new owner. The same applies to any third party to the sale. Owners of property that is damaged after the airplane is sold could also proceed against the builder if allegations are made that the airplane itself was improperly built, or maintained, and those facts caused the damage.

The first tactic to use to protect the builder is to see a competent aviation lawyer, and prepare a written contract of sale. This contract can contain some or all of the following provisions to the builder's benefit. Without a written contract, there isn't much protection offered to the seller, except hope. At a minimum, your contract of sale should have the following provisions:

- A statement that the aircraft is being sold "as is," with a disclaimer of all express and implied warranties, as your state allows, and with language that will suffice in that state
- An acknowledgment by the buyer that the aircraft is amateur-built, or restored by an amateur, as the case may be
- A statement that the buyer has had the aircraft inspected by maintenance personnel of the buyer's choosing, or that the buyer freely chooses to conduct his own inspection, and is relying solely upon that inspection as to the condition and safety of the aircraft
- A provision whereby both seller and buyer agree to the contract's being interpreted under the laws of a specified state
- An indemnity provision stating that the buyer agrees to hold the seller harmless from any subsequent litigation, and that the buyer will pay the seller's attorney fees and costs of litigation if the seller is subsequently

sued by anyone in connection with the buyer's operation of the aircraft, or subsequently relating to future owners, if the buyer later sells the homebuilt or restored aircraft

These special provisions should be included in addition to the normal contract paragraphs which cover such subjects as conveyance of good title, price, terms of payment, and place of delivery. A few dollars spent at this stage may pay off at many times the investment in the case of trouble down the road.

Buying a homebuilt can be very satisfying, or terribly frustrating, depending upon your own level of knowledge, and how well you do your homework. The obvious question about any homebuilt is how well was it built. The level of craftsmanship employed in the building is the most important query that a potential buyer must answer.

People buy a homebuilt for many reasons, the most common of which is that the buyer realizes that she doesn't have the time, inclination, or basic skills to build an airplane, yet she wants a particular kind of amateur-built aircraft. Relatively new homebuilts can be purchased readily now that homebuilding has become so popular. Most homebuilts represent quite a good value, in terms of the performance of the airplane for the dollars spent.

If you consider a homebuilt, remember what you're buying: someone else's product. Unless you have the knowledge and skills of a certificated mechanic, get an extremely thorough prepurchase inspection. Require the seller to produce the builder's log, and building photographs for inspection by your mechanic. Ask the builder to be available for any questions your mechanic may have. Then follow your mechanic's advice. Certificated maintenance personnel can spot shoddy construction or shortcuts that may make the airplane undesirable, or downright dangerous.

Also realize that maintaining a homebuilt that you buy won't be any less expensive than keeping a factory-built airplane in good condition. You can't be your own mechanic to any greater extent than that allowed as preventive maintenance to any owner-pilot. Unless you buy a very new design, airframe parts may be very hard to come by, even if a kit manufacturer will sell them to a secondhand buyer. Figure that you'll probably have to have many parts fabricated by the mechanic who maintains the airplane, and that can get expensive.

Homebuilding offers many advantages over buying factory-built airplanes. The number of kits being sold and assembled these days speaks loudly to that statement. If you want to get into homebuilding, study the rules, choose your project carefully, obey the regulations, and be prepared for years of work. Most homebuilders who stick it out and finish their aircraft wouldn't trade the experience for anything. They may sell their airplanes at some point in the future, but not generally for a long time.

Everything we do has some level of risk associated with it. Building one's own airplane doesn't have to entail any more risk than you're willing to accept, unless you overstep your levels of ability, knowledge, and available time. Built right, a homebuilt can be a very satisfying airplane for years. Done poorly, it can kill the pilot on the first flight. Know what you're getting into, get good advice and help when you need it, and your experience in the world of amateur-built aircraft will be pleasant and rewarding.

5

Aviation Insurance

INSURANCE IS A SUBJECT THAT IS CONFUSING TO MOST INDIVIDUALS who are not in some manner intimately connected with that industry. Very few people have any real idea of the sorts of coverages that their various insurance policies provide, the conditions under which the coverage is effective or void, and how to assure themselves that they are properly covered for the risks that they undertake. Unfortunately, these comments are not restricted to aviation insurance. Despite the fact that almost everyone has basic automobile and homeowner insurance coverage, it is surprising to realize how many folks truly do not understand their policies.

Even a reasonable degree of familiarity with the common coverages afforded in homeowner and auto insurance does not have much transferability to the subject of aviation insurance. In today's legal environment, proper insurance is an absolute necessity. Those engaged in professional occupations regard malpractice coverage and its attendant expense as a cost of doing business, just the same as office rent and support staff costs. All of us must be properly covered for automobile liability, and most of us who own our homes realize that homeowner insurance is as necessary as the roofs over our heads.

Aviation insurance can also be a mystifying topic. As with many of the other aspects of aviation, a good part of the philosophy and terminology used in aviation insurance had its beginnings in marine insurance. We still refer to the portion of the policy coverage that pays for the actual damage to an airplane as *hull coverage,* a phrase that is most apparently derived from the centuries of shipping insurance, when the hull of a ship was insured. Therefore, let us begin a discussion of aviation insurance with some of the peculiarities that are probably not familiar to most people.

USE RESTRICTIONS

When an underwriter at an insurance company evaluates the risk for which he is being asked to produce a premium quote, many factors must be considered. The type of airplane or airplanes involved in the proposed insured's operations, the qualifications of the pilots who will be flying them, and the use to which those aircraft will be put all go into the decision about whether a company even wants to insure that risk and, if so, for what premium and to what levels of coverage.

The first of these factors considered is the use to which the aircraft will be put. I will limit discussion to some degree to the lone-airplane situation, although, as necessity dictates, I will expand the thoughts contained here to fixed-base operators, flying clubs, and other multiairplane operators. The use restrictions are denominated in a policy of aviation insurance in three general categories: first, *pleasure and business;* second, *industrial aid*; and third, *commercial coverage.* Some insurance companies have another category of use, which is called limited commercial.

Pleasure and Business Coverage

By far the majority of the aviation insurance policies issued are limited to what is known as pleasure and business use. Pleasure and business, by the very terms, contain two anticipated uses of the airplanes. The first use, pleasure, is fairly obvious. This covers the use of the airplane when it is for the personal pleasure of the pilot who is permitted to be in command of the airplane.

The business part of pleasure and business coverage means use of the airplane in furtherance of the personal business of the owner or permitted pilot, such as a flight by a business owner to see a customer or supplier in another city. Pleasure and business use does not by any means include any use of the airplane for which a charge is made or compensation is paid to either the pilot, or for the use of the airplane for a particular flight. Pleasure and business use is totally noncommercial in nature; this type of coverage is applicable to the nonremunerative use of an airplane by the owner or other permitted pilot for her own recreation or travel for her own purposes.

Except as allowed by the provisions of one company's policy, under no circumstances may an aircraft be rented to another person or entity and still have a pleasure and business policy affording coverage in place for any flight when such a charge is made. Many an aircraft owner thinks that he can casually rent his airplane, on an occasional basis, to a friend or other person and still have his pleasure and business policy in force and in place, affording the coverage he thinks he has. Nothing could be further from the truth. In fact, a great number of the policies issued by aviation insurance companies for pleasure and business use prohibit even reimbursement for the use of the airplane for anything beyond the bare minimum of fuel and oil consumed during a particular flight.

If you happen to fly your airplane on the business of your own company, or for your employer, and you receive a reimbursement from clients, the employer, customers, or any other persons for the hours flown for that business purpose, it is imperative that you get a special endorsement on a pleasure and business policy to permit reimbursement. Should an accident occur during a flight for which you received reimbursement for any economic expense of the ownership or operation of your aircraft that exceeded the cost of only fuel and oil, you are at serious risk of having no coverage. In the event that you have a pleasure and business policy, a very careful review of your policy provisions needs to be made by yourself, your attorney, and your aviation insurance agent or broker to be absolutely certain that any reimbursed flight will be afforded coverage.

Industrial Aid Coverage

Industrial aid policies are issued to operators who fly their aircraft in the furtherance of commercial activities not connected with air transportation, aircraft rental, or training, such as those conducted by corporate flight departments. Very few individual aircraft owners, co-ownership arrangements, or flying clubs will come into contact with the industrial aid policy.

This policy is most often issued to operators of large aircraft within flight departments of corporations where the airplane or airplanes are used to carry customers, prospects, or employees of the business that operates the flight department. In fact, it is a type of policy that is generally used in the corporate world, and almost never used for the operator of small equipment.

Commercial Coverage

The full commercial policy is the type of coverage that is necessary for any operator who intends to be compensated for the use of his airplane. Therefore, fixed-base operators who rent airplanes—usually flying clubs, flight schools, and similar operations—must be certain that they obtain full commercial coverage as the permitted use of the airplanes to be insured under their aviation insurance policy.

Commercial coverage allows rental of airplanes to certain pilots—who those pilots are and under what conditions certain people are permitted to fly the airplanes will be discussed in a few moments. Commercial policies are generally tailored to meet the needs of the operator. Perhaps there is a fixed-base operator who rents airplanes, but does not have a flight school. Perhaps there is a flight school that conducts dual instruction and the appropriate amount of solo flight for students within its training curriculum, but that does not then rent airplanes to the general public.

The commercial policy will have several subcategories of protection available as well. Most fixed-base operators (FBOs) will have some kind of *premises liability coverage* to cover the liability to which they are exposed for accidents stemming from the condition of the premises they occupy. Chuckholes in runways, piles of plowed snow, wires, and similar hazards have nothing to do with the actual operation of that operator's airplanes, but still represent areas of liability exposure for which the operation needs insurance.

Limited Commercial Coverage

Limited commercial coverage is a type of commercial coverage issued by some aviation insurance companies to FBOs. This type of coverage extends generally to aircraft rental and flight training operations, and excludes the other types of commercial uses that are insured under full commercial policies. If the FBO has no charter flying, banner towing, aerial photography, or other commercial flights outside of rental and flight training, that FBO might well have limited commercial uses authorized in its insurance coverage.

Also, most fixed-base operators will want *hangar keepers' liability coverage* to protect themselves against liability having to do with aircraft that are hangared or otherwise stored at their facilities. Should a fire erupt, should one of the operator's airplanes taxi into a customer's airplane, should a hanger be destroyed in a windstorm, or should similar catastrophes strike that affect the aircraft stored on the premises, hangar keepers' liability insurance coverage will be available to protect the operator against the liabilities emanating from these losses.

Fixed-base operators who perform maintenance and who sell parts or aircraft will want to purchase two additional levels of coverage known as *completed-operations* and *product-liability coverages*. Completed-operations coverage generally is a provision within the policy that affords coverage for maintenance and overhaul functions, should the same be done negligently, or otherwise in a manner which causes loss to the owner of the aircraft and for which the fixed-base operator is then exposed.

If the fixed-base operator is involved in selling anything, he will certainly want product liability coverage to protect himself against the liability involved with placing products in the stream of commerce. Even the sale of fuel is the sale of a product, and should a fixed-base operator sell contaminated fuel, he will certainly want to have product-liability coverage to protect him in the event of an accident that might be caused by an aircraft's using that fuel.

In the last analysis, it is most important that each aircraft owner and pilot be absolutely certain that he does not violate the use restrictions contained in his policy, or that he contact his broker or direct-writing insurance company and change the permitted use to correspond with the application to which he puts his airplane. Failure to do so will likely void the coverage that otherwise would be afforded by the policy, and will render the owner uninsured.

There is probably no worse result than for a pilot or owner to think she is insured when she is not. Some people make the judgment call to go without insurance, not only in aviation but in other aspects of business and life in general. Those decisions are at least made before the activity is undertaken, and the person who is choosing to fly without insurance is not operating under a delusion that coverage exists.

But it is another situation entirely to be aviating along, violating basic policy provisions that will result in the company's denying coverage should an accident occur. Simply do not let yourself fall into that trap. Know what coverage you have in terms of what use to which you are permitted to put the airplane, for there is no better protection than a thorough knowledge of all the terms of your policy, particularly in this area.

PILOT WARRANTIES

Because of the extremely wide range of performance characteristics within the general aviation fleet, and a similar wide range of pilot skill and experience, an aviation insurance company will undoubtedly require that pilots who fly the airplane meet certain conditions. The phrase *pilot warranty* is used because the courts have interpreted this particular portion of the policy as a warranty that is being given by the insured to the insurance company that only those pilots who meet the specified conditions will fly the airplane.

One of the best ways for an aviation insurance company to minimize its exposure is to require that the pilots who fly the airplane be competent to do so. All policies will require that a pilot be properly licensed and certificated to operate the aircraft in question. There are virtually volumes of cases and court decisions concerning situations wherein an insurance company has attempted to deny coverage for an accident because the pilot did not meet the specified requirements. In this situation, the company will allege that the insured breached her pilot warranty and did not provide a pilot for that flight who was qualified, under the provisions of the policy, to be pilot in command.

In the area of pilot warranty, remember that there is more to being properly qualified and certificated for the flight beyond simply holding a pilot certificate issued under Part 61 of the FAR. There are requirements for biennial flight reviews, currency in classes of aircraft, and instrument currency, and a pilot must generally meet all these requirements in addition to the simple possession of a pilot certificate to be properly qualified and certificated under the policy terms. If you fly outside your biennial flight review date, if you do so without a valid medical certificate, or if you fly an aircraft in which you are not current, you are not a properly qualified and certificated pilot, according to most courts in most states. The same applies for an instrument pilot who flies IFR when he does not meet the instrument currency requirements of Part 61.

Also, remember that even though a pilot possesses a medical certificate, that medical certificate is deemed by Part 67 of the regulations to be invalid during a period in which the pilot suffers from a physical deficiency, of which he knows, which would render him unqualified for the issuance of that medical certificate. So even if a pilot has a medical certificate that, on its face, appears to be valid, if he is operating during a period of known physical deficiency, the insurance company may well be able to successfully deny coverage in the event of an accident.

When a pilot fails to meet the requirements of the pilot warranty, and has an accident, the insurance company will most often seek to deny coverage for the accident. Some states require that the company prove that the lack of the pilot's qualifications was the cause of the accident before the courts in those states will void the insurance coverage.

Other states do not require such a showing of a causal connection between the pilot's lack of meeting the pilot warranty and the accident itself. Ohio is one of these states. An accident occurred some years ago in the Cleveland area in which a Beech Bonanza suffered airframe damage in a gear-up landing that was caused by a mechanical failure of the landing gear mechanism, not the pilot's forgetfulness. The insurance policy required the pilot to have 15 hours in a Bonanza. Even though the pilot had lots of time in many different aircraft with retractable gear, and was properly certificated as a pilot and had a valid medical certificate, he was about 3 hours short of the required 15 hours in a Bonanza when disaster struck.

The insurance company involved filed a lawsuit to declare that the insurance coverage of the policy was not in force at the time of the accident. Even though the pilot was able to show that he was a competent and experienced pilot and, further, that the cause of the accident had nothing to do at all with his piloting technique, the insurance company won the case, and the owners of the airplane had no insurance money to fix their Bonanza's scraped belly. The court held that a pilot warranty is just that—a warranty by the insured to the company. When the insured violated the warranty by permitting the aircraft to be flown by a pilot who did not meet the experience requirements of the warranty, the company was free to deny coverage for the accident. The moral of the story is clear—meet your pilot warranty, or have no coverage.

Beyond requiring the pilot to possess a pilot certificate and otherwise be qualified to fly the aircraft, most pilot warranty clauses will demand minimum levels of certificates, minimum total experience, experience in the class of aircraft involved, and probably experience in the very type.

For most light aircraft, especially those with fixed landing gear and fixed-pitch propellers, most insurance companies will require in the pilot warranty that the pilot be any properly certificated and qualified private or commercial pilot. In this type of aircraft, few companies require an instrument rating. It is not difficult to get most companies to even broaden the pilot warranty to permit student pilots to fly this category of aircraft.

But once the complexity of the aircraft increases, so will the severity of the requirements of the pilot warranty. Most companies today will definitely look favorably upon the pilot of a complex airplane who has an instrument rating. There are some companies that simply will not insure complex aircraft to be flown by pilots lacking an instrument rating.

For aircraft with constant-speed propellers and retractable landing gear, it is most common for the insurance company's pilot warranty provisions to require that the pilot have a significant amount of time in retractable-gear aircraft. The pilot will most likely be required to have some time in the specific type of aircraft insured before she meets the pilot warranty requirements.

One way to control your cost of insurance is to be judicious in the pilot warranty that you are willing to accept. In the situation in which most owner-pilots find themselves, that being where they and they alone fly their airplanes, there is no reason to have a pilot warranty more liberal than the circumstances require. If you possess an instrument rating and fly a Cessna 182, suggest to your insurance agent or direct-writing company representative that you would be willing to accept a pilot warranty that requires an instrument rating. Likewise, if you have an advanced certificate, make a similar suggestion.

Because the insurance market is very cyclical, there will be policy years when premiums will rise and others when premiums will fall. In the early to mid-1980s, when the insurance market was very tight, premiums rose dramatically, and some folks found they could not get the coverage they had before. Then, by the 1990s, things had settled back down. Premium rates fell, and coverage was easier to obtain. When a tight situation strikes again, and it invariably will, some creative negotiation with your company may put you in a position to obtain insurance that you otherwise might not obtain, or bring down the cost of it. One of those areas of creativity is the requirement of your particular pilot warranty.

HULL COVERAGES

The policy provisions that obligate the insurance company to compensate the owner of the aircraft for physical damage to the aircraft itself are the items in the policy known as *hull coverage*. As previously stated, this phrase came from the custom in marine insurance of insuring the hull of the ship in addition to the cargo contained in it. As applied to aviation insurance, the word *hull* simply means the airplane itself and its engine, propeller, accessories, and avionics.

Hull insurance is very similar to collision coverage in the auto insurance industry. It is basically a no-fault type of coverage, which means that if the aircraft is damaged by whatever means, the owner is entitled to coverage and payment on the loss, provided that the use to which the airplane was being put and the pilot flying the airplane met the policy requirements.

Aviation insurance is one unique industry in which most companies will issue policies that contain *stated-value* coverages for the hull provisions. This means that when the insured is in the process of purchasing the coverage, he tells the company the amount for which he wants his hull insured. Then, in the event of a total loss, that is the amount paid. A contrary system exists in auto coverage, where most policies simply obligate the insurance company to compensate the owner for the fair market value of the vehicle at the time of a total loss. In auto insurance, it is not at all uncommon to get in a dispute with the company over the value of the car when it is destroyed.

But in aviation insurance, such disputes are rare. Because the company and the insured agree at the time of policy inception as to the stated value of the aircraft, if it is totally destroyed, both parties to the policy of insurance know precisely what coverage is available, and are further aware of the amount to be paid for that total loss.

The subject of what constitutes a *total loss* merits some brief discussion. An aircraft will be considered totally destroyed, for insurance purposes, when the cost of repairing it after an accident exceeds the stated value for total loss, less the amount of salvage value that the company can recoup when it sells the wreckage. When an aircraft is totally destroyed, and the company pays the stated value for total loss, the company is entitled by the policy provisions and by the common law in most states to obtain title to and possession of the wreckage. In most situations, that wreckage has some salvage value. Obviously, there are those accidents in which the wreckage is so mangled that it has little or no value; but in the majority of accidents, a substantial salvage will be obtainable from what remains of the aircraft.

Therefore, an owner can easily outsmart himself if he does not insure his aircraft for enough. Take the example of an aircraft that has a true market value of $30,000, but a pseudointellectual owner decides that he will insure it only for a $20,000 stated-value hull coverage. Then, in the event that an accident occurs where the repairs would cost the company $21,000, the company may well total the airplane and pay off its stated-value obligation ($20,000), and take title to an airplane that is well repairable.

The company would undoubtedly sell the airplane to an aviation salvage operation that would, in turn, probably repair and sell it, or maybe part it out after disassembly. Our owner would have suffered the short end of that bargain by being penny wise and pound foolish. Therefore, when determining the amount of stated-value hull coverage for which you wish to insure your airplane, be reasonably certain that the stated-value coverage is fairly close to the actual market value of your pride and joy. To insure it for less might result in the loss of your airplane in a situation where, had you purchased enough coverage, it would be repaired.

Conversely, an aircraft owner should not try to overinsure an aircraft. Most insurance companies are quite liberal in this regard, as everyone thinks that his airplane is a perfect specimen of the type and worth more than the general marketplace would dictate. Because insurance is one of the most people-oriented and emotionally sensitive industries going, companies understand that most aircraft owners probably do think their airplanes are worth somewhat more than the free market would dictate, and accept slight overevaluations in stated-value coverage; and when a total loss occurs, they willingly pay those slightly inflated values.

But an aircraft owner cannot grossly overestimate the value of her aircraft for insurance purposes. Some policies expressly provide for that eventuality and permit the company to pay the stated value or the fair market value, whichever is less, in the event of overinsurance of the hull value. Even in situations where the policies do not contain such provisions, the law of many states will do it for them. We must all remember that one of the basic precepts of our legal system is that of fundamental fairness. Because insurance fraud is a very sensitive subject, and one to which all insurance companies, regardless of the specific niche of the insurance industry in which they do business, are extremely sensitive, if one grossly overinsures the value of an airplane, expect litigation to result if the insured tries to insist upon payment of that grossly overvalued sum.

The amount for which the hull is insured needs to be determined in consultation with an experienced aviation insurance agent or broker, or direct-writing company representative. These individuals generally have a very good handle on the marketplace and can very closely estimate the real value of most common aircraft. The decades of the 1980s and 1990s have seen tremendous increases in the value of many general aviation airplanes. If you've owned your airplane for a few years, and have not kept abreast of its value, you may be in for a surprise when you do your homework. But it is better for that surprise to be a pleasant one, when you discover your airplane's present worth in discussing it with your insurance broker or agent at the time that you are purchasing coverage, rather than being in for a rude awakening after your airplane is totaled, the company pays the stated value in the policy, and you're far short of the amount of money needed to replace it. Little is to be gained in deviating from the conventional wisdom by attempting to either excessively undervalue or overvalue the coverage of the hull.

LIABILITY COVERAGES

Chapters 7 and 8 cover the principles of liability that the law has developed to give us a system whereby those who suffer loss at the hands of another can be compensated. At this stage, we are not particularly concerned with why an individual or business entity may be subjected to liability, but here we shall examine the types of *liability coverages* that are generally available in the aviation insurance marketplace. All policies of aviation insurance will have a maximum limit of liability protection.

That maximum limit is often stated in multiples of $1 million, and the uninformed will assume that they have coverage in that large amount. For most accidents, and under the terms of most policies, that is simply not the case, and I would venture to say that the majority of the pilots flying today do not realize the true extent of their liability insurance protection.

Per Person Limitations

Unless the insurance policy that an aircraft owner has is of the combined single limit type to be discussed later in this chapter, that policy will have *limits of coverage that will apply to each person or passenger* who is injured or killed in an accident. This type of coverage provides that when a person, sometimes excluding passengers in the airplane, suffers injury or death as a result of some act or omission of the pilot for which the insured is legally liable, the insurance company agrees in its policy to pay that injured person, or her estate if she is deceased, for the liability of the insured, up to certain limits.

These per person limits have to be considered in conjunction with the total amount of liability protection that the insured gains by virtue of his policy. It is very common to see aviation insurance policies wherein the total limit of liability is $1 million, but those same policies go on to provide for limits per person, or per passenger, of $100,000 and occasionally $250,000. What this means is simple, once it is analyzed. For each person or passenger who is killed or injured in the accident, the insured has protection available under the policy to the amount stated as the per person or per passenger limit. If that limit is $100,000, there is only $100,000 worth of coverage available for each injured or deceased person or passenger.

If you fly a four-seat airplane, which obviously therefore is capable of carrying only three passengers in addition to the pilot, and if you have a policy offering total limits of $1 million and $100,000 per person, per passenger, or per seat, it becomes readily apparent that there is available the sum of $100,000 each for the injuries or death suffered upon each of those passengers. Therefore, the $1 million of total liability, while remotely possible to exhaust under some circumstances, is not a real figure in most cases. What becomes most important is the amount of coverage available to each person, including passengers, who can suffer injury or death in such an accident.

In order to exhaust the total limits of liability in our hypothetical scenario of a $1 million total policy with $100,000 per person limitations, we would obviously have to cause injury or death to 10 persons to the extent that each would be legally entitled to at least $100,000, or cause a large amount of damage to the property of others. For most of us flying small, four-place aircraft, unless the accident involves the aircraft's going into a building and doing catastrophic damage, the company's real exposure to pay the $1 million total limit is quite remote.

Therefore, each insured needs to view his policy in terms of the per person limitations that it contains. Per accident limitations (the maximum liability limits) are important, but far less so than the per person limitations commonly used by the aviation insurance industry.

Each insured must examine his own type of flying and his own exposure. If most of your flying is solo, perhaps you can live with a per person limitation of $100,000, which is currently the common limit in most light-aircraft pleasure and business policies. If, to the contrary, most of your flights involve carrying passengers, particularly high-income individuals whose injury or death would result in large liability claims, $100,000 per person would be wholly inadequate to compensate either them or their estates in the event of the unfortunate. Again, proper consultation with your broker or direct-writing company representative, combined with legal advice from a competent attorney who practices in the liability area, is essential to determine the proper extent of coverage that any particular aircraft owner or pilot should purchase.

Combined Single Limit

The aviation insurance industry constantly modernizes its policy forms and the types of coverage offered; and several years ago a new form of liability limit came into vogue. The *combined single limit* of liability is a provision in the policy that allows for the entire, gross amount of the policy coverage to be available for the property damage and bodily injury or death of any individuals that might result from an accident. Unlike the per person limits, the gross amount of the liability limit contained in a combined single limit policy is available to satisfy the claims of those people who suffer such an injury in an accident.

Again, some elementary mathematics must come into play in order to analyze the concept of a combined single limit. If you purchase, as part of your total aviation insurance package, a combined single limit of $1 million, you then have $1 million available to satisfy the claims of all those who suffer damage as a result of an accident. If the accident occurs while the aircraft is carrying one passenger, and that is the only individual who suffers any legally recognizable injury, the full $1 million is available to satisfy that person's claims. If two passengers are carried, the $1 million must be, in some manner or another, split between them. That does not mean that the split will be 50-50, since one passenger might be injured significantly more than another, and therefore present a much larger claim than the other. The point is that there is $1 million available to satisfy the claims of all those who suffer injury as a result of the accident, whether they be one in number or several.

I prefer the combined single limit approach, and almost always recommend it to clients. It is a commonly known fact throughout the aviation field that it is a very rare flight in general aviation aircraft when all the seats in the aircraft are filled. Most flights involving the light, single-engine aircraft carry one or two passengers at the most. So the combined single limit approach to liability insurance affords more real protection than does the older style of per person limits, because the gross amount of the policy's limit of liability is available to satisfy what will, in all likelihood, be claims by a relatively small number of injured persons.

Because with a combined single limit approach the insurance company's exposure is much greater, there is almost always a premium difference between combined single limit coverage and coverage with per person limitations. In my view, the combined single limit of liability is well worth what is usually a very small increase in premium.

If the aircraft involved is something unique such as an antique, a homebuilt, or a warbird, you might not be able to find an insurance company that will offer combined single limit coverage. Even if the aircraft is of normal category, if it is used for something that most companies consider more hazardous such as pipeline patrol, parachute jumping, or a similar endeavor, most companies will not offer combined single limit policies. But for the kinds of flights that most of us take, with average pilots, flying normal-category airplanes, there are very few cases that do not significantly benefit from a combined single limit policy.

Renter Pilot Coverage

Many general aviation pilots do not own their own aircraft and never will. The reasons behind this are many, and often revolve as much around the fact that the average pilot flies only 50 to 60 hours per year as around the financial limitations put upon all our dreams. When a pilot flies fewer than approximately 100 hours per year, the costs involved do not justify ownership of an aircraft. As stated in Chapter 3, that burden can be lessened somewhat through one of the various co-ownership arrangements, but cannot be eliminated.

Renter insurance, more properly known as *nonowner insurance*, is now offered by a large number of aviation insurance companies because, where there is a market, wise suppliers will step in. Many pilots rent their aircraft from a fixed-base operator, maintain some kind of membership in a large flying club where perhaps they do not actually own an equity interest in the aircraft, or even borrow a friend's aircraft. Obviously the most common approach taken by the nonowner pilot is to rent aircraft from the local fixed-base operator.

There is a cliché in aviation that the way to make a small fortune in this industry is to start with a large one. There is a lot of truth in that statement, and when a person becomes familiar with the costs, overhead, and thin profit margins involved in most fixed-base operations, shock sets in. Therefore, most every fixed-base operator is extremely cost-conscious, and one of the largest pieces of overhead that must go into the FBO budget is the premium to be paid for insurance. Because the expense of insurance is so large, fixed-base operators must constantly be searching for ways to reduce that cost. One of those ways is to purchase aviation insurance that does not offer protection to the renter pilots to whom the aircraft are supplied, but rather protects only the liability of the fixed-base operator itself.

I have talked with many nonowner pilots who never dreamed in a thousand years that they were not covered by the insurance purchased by the fixed-base operator from which they routinely rented aircraft. A good number of states, but still a minority, have laws on the books that require those who rent airplanes to disclose to their customers the limits and extent of insurance provided by the fixed-base operator or other lessor. All states should have it, but most do not.

So the nonowner pilot is in a position where, in most areas, there is no requirement that she be informed of the insurance provided by the person from whom she is renting. Even when the question is directly asked, many FBO personnel will have a standard line concerning the limits of liability of the policies that they are authorized to divulge to customers, but the disclosure does not include the fact that the pilot himself is not covered. If you are a nonowner pilot who rents aircraft, you need to know whether you are covered under the FBO's policy, or whether that policy simply provides protection for the FBO's liability and forces the renter pilot to fend for himself.

Because the lack of renter pilot coverage is so prevalent in the FBO industry, all renter pilots should definitely consider the purchase of their own nonowner liability coverage. Even if the FBO's insurance extends to the protection of the renter pilot, the renter is liable for any loss that is not insured. If the FBO has missed a premium payment or in some other manner violated the provisions of the insurance policy, that operator and its renter customers can all be without insurance protection.

Personally, I do not like my fortunes riding on whether or not someone else paid his premiums, had a proper pilot warranty, and insured his aircraft for the use to which he is actually putting them, or for the amount of hull coverage that is reasonable. Renter pilots have a great deal of exposure, most of which they do not realize, and only a nonowner liability policy of appropriate limits gives them the protection that they need.

Also, the FBO may have extremely low limits of liability in its policy. Even if coverage does extend to the renter pilot, the pilot might be legally at fault in an accident for which the FBO's insurance company steps in, quickly pays its limits to the injured person or the estate of a deceased person, and gets out. Then, the renter pilot who actually caused the accident is once again left adrift in a sea of liability with no insurance protection.

So all renter pilots should purchase *liability protection;* and at small levels of limits, such policies are often available for premiums in the neighborhood of $250 annually, or less. Most often, renter policies are not available in the larger limits of liability that companies sell to aircraft owners. But some protection is still better than none.

Renter pilots should also consider purchasing, as part of their nonowner policy, *coverage for aircraft physical damage.* This coverage provides protection for any amount that the renter might be compelled to pay for physical damage done to the aircraft, even though no person suffers bodily injury. The FBO will undoubtedly have a deductible on its hull insurance, as will a private owner of an aircraft. Then if an accident occurs, the renter pilot, if she caused the accident, is liable for that deductible amount.

Furthermore, there are several FBOs that simply do not carry hull insurance on their aircraft at all. Many of them are dealers or distributors, or are in another situation where they can purchase parts at wholesale or distributor prices and have their own maintenance facilities, thereby making their cost of performing repairs less than the retail labor costs. Such FBOs have made a business decision that they will not pay the premiums for hull insurance, and take the risk of repairing physical damage themselves when their aircraft are involved in accidents. The wisdom of such practice is obviously open to debate, but many longstanding and successful FBOs follow the practice of not purchasing hull insurance for their aircraft. If you happen to rent from such an FBO, you are potentially liable for the full cost of repair to the aircraft for any damage that you cause and for which you are legally liable.

Nonowner insurance is generally available only for pleasure and business use. There is little or no such coverage available for commercial, charter, or other special-use purposes. Likewise, territorial limitations need to be considered. Most aviation insurance, including but not limited to nonowner policies, will be effective only in the United States, Canada, and in some instances parts of the Caribbean. If you fly into Mexico, make certain that you get insurance that is effective there, since Mexican law is significantly different from American and Canadian law, and the insurance issued in the United States is often not recognized or valid in Mexico. A similar warning needs to be considered if you fly into any of the islands of the Caribbean or the Bahamas. All aviation insurance does have territorial limits. Do not simply purchase your policy, put it in your filing cabinet, and then venture off to foreign soils—perhaps Europe via the North Atlantic—assuming that you are covered. You might not be, unless that foreign soil is Canada.

DUTY TO DEFEND

A part of almost all casualty insurance—which includes aviation, personal liability, and auto insurance—is the duty of the insurance company to provide the *legal defense* for its insured when the insured is sued over an event that is covered in the policy.

This means that even if the claim of the plaintiff is without merit, and victory is almost assured for the defendant, the insurance company must come forward and provide a legal defense. Because the company has the duty to provide and pay for the defense, it has the right, both under the policies that are issued and under the common law as well, to choose the defense counsel to be employed. There are some exceptions to this general rule, as there are to most general rules, but unless there is a coverage dispute pending or some other special circumstances, the company will choose the attorney or law firm to defend the insured.

The value of the duty to defend cannot be understated. Even the most careful pilot, and the most skilled, can be sued by people who present claims or who hope to accomplish a quick settlement, thinking that the defendant would rather pay off quickly than undergo the trauma, emotion, and expense of protracted litigation.

The costs of a proper defense in an aviation accident case can be astounding. Because of the complexity of these suits, not only are legal services required, but in most instances the defense must employ one or more expert witnesses to assist in the preparation of the case, opine as to the various issues of causation, and explain them in live testimony to the court and jury.

It is not at all unlikely that $200,000 or more can be spent in the process of defending a serious aviation case. If you choose to go uninsured, think about the costs involved in a successful defense. Another cliché that has been used in certain professions is that the operation was a success, but the patient died. While not intended to be humorous, such an analogy can be drawn to the defendant who finds himself in the unfortunate position of defending a case that he eventually wins, but suffers financial disaster in the process. Worse, he might not be able to find competent counsel simply because he is unable to fund a proper defense.

This duty to defend should be enough motivation to cause a renter pilot to purchase nonowner liability insurance. But if that is not enough, couple the cost of defense with the possibility that, in addition, a substantial settlement or judgment might have to be paid, and it is plain that every person who rents or otherwise uses aircraft belonging to another should have adequate nonowner insurance.

If you own your own airplane, the same arguments apply. The insurance company has the same duty to defend even worthless claims, let alone the real ones. If you fly without insurance, either as an aircraft owner or as a renter pilot, you risk losing all that many people have in defending yourself in a lawsuit, even if you win it. If you lose the case, or need to settle with the plaintiff short of trial, a bad situation only gets worse.

SUBROGATION

Subrogation is the term used to describe the legal right that an insurance company possesses, when it pays its insured for a covered loss, to step into the shoes of the insured and seek recovery from the person who caused the damage. This concept usually arises when a hull insurance claim is paid for the damage to the aircraft caused by someone else. Perhaps you land your airplane at an airport where snow has recently been plowed and piled along the runway too close to allow safe clearance between the wings and the snow piles. The landing occurs at night, and damage occurs to your airplane when the pilot, through no fault of her own, strikes the snow piles.

In this case the insurance company will pay the insured for the repair to its airplane, or perhaps for the total loss of it, but then will step into the shoes of its insured and succeed to the right that the insured would have had to sue the person or business entity that caused the problem in the first place. In such a hypothetical situation, the insurance company could then file suit against the airport, or fixed-base operator, or any other entity or person who negligently and improperly plowed the snow and piled it too close to the runway.

Another type of subrogation can arise when loss occurs to the aircraft of a fixed-base operator, flying club, or other renter of aircraft. If the lessor of that aircraft does not have the extended coverage on its policy that protects its renter pilots, the renter pilots are strangers to the policy between the aircraft lessor and its insurance company. Therefore, if the renter pilot causes an accident through his own fault, the insurance company will step in, pay the lessor for the damage to its aircraft, and then have subrogation rights to sue the renter pilot whose fault resulted in the damage for which the company has paid.

There is a legal principle that an insurance company cannot subrogate against its own insured, so if the FBO or other aircraft lessor has the coverage that extends the protection of the FBO policy to its renter pilots, the renter pilots generally do not have to worry about subrogation. However, if the FBO does not have this extended type of coverage, as discussed above, the renter pilot needs to be significantly concerned that if he damages the airplane that he rents, even though no personal injury occurs he could still face the requirement to fully reimburse the FBO's insurance company for the damage caused through his fault. Such is not an attractive prospect for renter pilots, and should be further motivation to purchase nonowned insurance coverage.

ILLEGAL USE OF THE AIRCRAFT

All policies of aviation insurance that cover the normal general aviation uses that we have discussed at the beginning of this chapter contain a policy exclusion, wherein *coverage will be completely denied for any use of the aircraft that is illegal.*

The wording varies in policies as issued from company to company. Some policies require that the insured actually have knowledge of the illegal use, while other policies seem to imply that, as long as the insured has knowledge of the flight, she does not have to have knowledge of the actual illegal use for the exclusion to become operative. The reason that the owner is required, at a minimum, to have knowledge of the flight, is that it is obviously illegal to steal an aircraft. Therefore if the policy simply excluded all illegal uses, the moronic result would be that a theft of the aircraft would not be a covered event. Obviously no company intends to exclude a theft, and thefts are routinely covered and paid under the hull loss provisions.

Each policy must be read carefully to determine the exact character of the illegal-use exclusion. If a contract of insurance has such an exclusion that requires the insured only to have knowledge of the flight, but not knowledge of the illegal use, the insured might end up in a situation wherein, in good faith, he lends his aircraft to a pilot who meets the pilot warranty; that pilot then proceeds to engage in an illegal use of the aircraft, of which the owner has no foreknowledge. Should an accident occur on such a flight, the insured can find himself with no coverage, either hull or liability.

If the policy in question requires the insured to have not only knowledge of the flight but actual knowledge of the illegal use, and the insured can prove that he did not have knowledge of the particular use to which the pilot was going to put the airplane, that insured will probably prevail. But anytime an aircraft is involved in an illegal use and an accident ensues, the insured can rest assured that the company will scrutinize coverage extremely closely and will attempt to deny coverage for the mishap if at all possible. Therefore, whenever an aircraft owner chooses to allow someone else to fly his airplane—even if that aircraft owner is a noncommercial user who lends his aircraft, within the provisions of the pleasure and business use, to a pilot who meets all the pilot warranty requirements—he is gambling that his friend will indeed remain a friend, and not proceed to put the aircraft to some use that the law prohibits.

Illegal uses can be many. The first that probably comes to mind to most of us is smuggling of drugs, other controlled substances, and other illegal contraband. In almost all situations, the carriage of those materials will be illegal and will result in the enforcement of the illegal-use exclusion in the insurance policy. Therefore, be extremely careful about the purposes of the flights that others might take in any aircraft you own, or in which you have an interest.

Let's take a look at a case that really happened a few years ago, and see what the result was. A fixed-base operator rented a Piper Arrow to one of his good rental customers for a 2-week cross-country flight to Florida. The pilot also intended to visit the Bahamas, and the FBO had the appropriate territorial limits in the insurance policy which permitted flights to that part of the world.

About 10 days after the customer left, the FBO received a phone call from the pilot. The renter was breathing heavily into the phone, obviously in a state of excitement. He related a story that he was coming back to Florida from the Bahamas, and was only a few miles from shore when the engine quit, and he was forced to ditch the airplane. He claimed that he was able to stay alive by using his styrofoam cooler as a floatation device, and spending a few hours swimming. He said no boats ever came in sight to pick him up. The FBO made a claim on his insurance policy for the hull coverage amount, since the airplane was obviously lost.

An investigator at the insurance company had a premonition that the pilot's story didn't add up, particularly the part about swimming for hours and seeing no boats near the Florida shoreline. For reasons that can't be totally revealed, a search was mounted to find the airplane. Low and behold, within a few days it was found in Jamaica. It was wrecked, and half burned, a few hundred yards from the end of a small airstrip owned by a plantation, and used legally by crop dusters.

Someone had installed a homemade long-range fuel system in the Arrow. It was also apparent that the airplane was burned from a fire that was set, since the fire consumed only the cabin portion, and there was only minimal fire damage under the cowling. The bad guys weren't very thorough; the registration number on the side of the fuselage was still very readable. It didn't take a genius to figure out that this was the end of a bungled drug run. The pilot hadn't studied the effects of density altitude and high gross weight enough to realize that an Arrow, full of its normal fuel supply, plus the extra gas in the long-range tanks, and loaded to the gills with wacky weed, wasn't going to get out of a short strip in the tropics.

The insurance company refused to pay the claim for the FBO's loss of the airplane, since it was being used for an illegal purpose at the time of the accident. The last I heard, the pilot, who had been a deputy sheriff, had changed his status from jailer to jailed.

Another illegal use is somewhat more subtle, but no less wrongful. That is the unauthorized furnishing of transportation by air for hire or, as most of us would call it, a charter flight. If you are simply a FAR Part 91 operator (as most general aviation owners and pilots are), neither you nor any person operating your airplane who does not have a Part 135 ATCO certificate, which allows charter flights—and one with your airplane listed on it—is authorized to furnish air transportation for hire, for the transport of either persons or property.

If the person who borrows your airplane, who you think is going to be operating within the confines of your insurance coverage, proceeds to violate those policy provisions and engage in a charter flight for hire, or uses the aircraft for hire in some other manner not covered by your policy, you can quickly expect once again that the company will deny coverage if an accident occurs.

6

FAA Enforcement Procedures

THE FEDERAL AVIATION ADMINISTRATION IS NOT ONLY CHARGED WITH formulating and promulgating the Federal Aviation Regulations that govern virtually all aspects of the aviation industry, but is also responsible for enforcing those same regulations. In all probability, the first violation of an FAR followed very shortly after the establishment of the regulatory scheme.

The attitude of the FAA toward enforcement is never fixed, and over the decades has taken some rather wild cyclical moves. In recent history there have been two primary enforcement crackdowns. The first began shortly after the air traffic controllers strike in 1981 and continued through the better part of 1983. Then, after the midair collision near Cerritos, California, between a Piper Cherokee and a Mexican airliner a few years ago, we saw probably the most vigorous enforcement period yet endured. That period lasted for several years. As of this writing, there is good reason to believe that this policy of strict enforcement might be lessened to some degree, owing in good part to the FAA's adopting a remedial training program to address regulatory violations that are the product of a pilot's ignorance, rather than actual misconduct. I'll have more to say about this program at the end of the chapter.

When a violation comes to the attention of the FAA, the agency basically has three choices of how to proceed. It can handle the matter administratively, which is little more than a warning; it can seek a civil penalty, which in common language is a fine; or it can attempt to suspend or revoke the particular certificate of the individual or business entity that allegedly committed the violation at issue.

Before taking any type of action, the FAA must get word of an alleged violation from some source. Quite frequently, this source is air traffic control, particularly in this day of Mode C transponders, which enable controllers to automatically know the altitude of aircraft as well as the geographic location. Controllers have little or no discretion under the current policy and must initiate enforcement activities by notifying the local flight standards district office (FSDO) of a deviation from clearances, airspace limits, landing or takeoff instructions, and the like. The FSDO often receives complaints from law enforcement officials, from the citizenry at large, and, less often, from other pilots or maintenance facilities. Regardless of the source of the complaint, the legal procedure is the same.

LETTER OF INVESTIGATION

Upon receiving information that a possible FAR violation has occurred, an inspector within an FSDO will be assigned to perform a very preliminary investigation into the allegation. A part of this initial approach is to send a letter, which is officially called a *letter of investigation*—nicknamed a 10-day letter for reasons that will shortly become obvious—to the certificate holder who appears to have violated the regulation. This certificate holder might be a pilot, a mechanic, a business entity that holds a Part 135 certificate or approved repair station certificate, or any other operator who is required by the FAR to possess some form of Federal Aviation Administration certificate.

The letter of investigation is carefully worded, and approved forms for such are contained in the official policy manual put forth by FAA's Washington headquarters. The letter begins by telling the recipient that a possible violation of the regulations has occurred, and that the particular FSDO is investigating. Very little specific information is given, and it is that way on purpose. The letter generally closes by informing the recipient that she has a period of 10 days in which to contact the writer, who is an FSDO inspector, and present her side of the story either orally or in writing. Figure 6-1 contains a sample of a letter of investigation.

Response to Letter of Investigation

This very carefully worded letter of investigation is calculated to accomplish a couple of purposes. First of all, naturally, it is putting the recipient on notice that an investigation is under way. But more important for the certificate holder, it seeks information by requesting a response within 10 days.

Through all the television shows seen over the last several years, people have become aware of certain rights of which law enforcement officers routinely advise persons about to be taken into custody and interrogated. These rights come from a United States Supreme Court case entitled *Miranda* v. *Arizona*, and are now ingrained as a part of constitutional criminal law. These rights include the right to remain silent, to have a lawyer present during any questioning, and to terminate questioning at any time the person chooses. These Miranda warnings are totally inapplicable to FAA enforcement actions. Miranda warnings are required to be given only to a detainee in a criminal case before any interrogation occurs, and because FAA enforcement actions are not criminal in nature (as the rules are presently written, there is no opportunity for the violator to be incarcerated), and because the recipient of a 10-day letter isn't in the custody of law enforcement officials, these warnings are not required. Even though the recipient of the letter of investigation is not informed of his right to remain silent, he certainly has such an option.

Therefore, the FAA inspector does not have to inform the recipient that what he says in his response to the letter of investigation can, and most probably will, be

July 5, 19--

File Number: 87CE040235

Mr. John D. Smith
1711 Colorado Avenue
River City, Iowa 51649

Dear Mr. Smith:

Personnel of this office are investigating an incident occurring on July 4, 19--, which involved the operation of Cessna aircraft N57785 in the vicinity of City Park at approximately 3:15 p.m.

The aircraft was observed and identified as Cessna N57785 diving on picnickers and bathers from 3:15 to 3:35 p.m. We were informed that Cessna N57785, piloted by you, landed at the airport at 3:45 p.m. Operation of this type is contrary to the Federal Aviation Regulations.

This letter is to inform you that this matter is under investigation by the Federal Aviation Administration. We would appreciate receiving any evidence or statements you might care to make regarding this matter within 10 days of receipt of this letter. Any discussion or written statements furnished by you will be given consideration in our investigation. If we do not hear from you within the specified time, our report will be processed without the benefit of your statement.

Sincerely,

JOHN L. DOE
Aviation Safety Inspector

ATTCH: PRIVACY ACT NOTICE

Fig. 6-1. Sample letter of investigation—flight operations.

used against him. Frequently the FAA lacks vital information that it will be hard-pressed to discover through its own endeavors. A classic example is one of an airplane operating VFR with no flight plan files whose pilot contacts ATC for clearance through Class B airspace. Perhaps the controller advises the pilot to remain outside the Class B, or gives the pilot specific heading and altitude restrictions with which she must comply while operating within the Class B. Regardless, assuming that the pilot either penetrates the Class B without a clearance or deviates from the clearance restrictions, the controller will then begin an enforcement action against the pilot of the airplane.

But how does the FAA know who the pilot is? Plainly, it does not. Yet when the letter of investigation is sent to the registered owner, the inspector might indeed be assuming that, in the case of a small aircraft operated on a Part 91 flight, the registered owner was in fact the pilot in command at the time.

If the owner was that pilot and responds to the letter of investigation and discusses the conduct of the flight in the first person, there is no doubt that the pilot has now provided the FAA with vital information which it would have had a very difficult time ascertaining without the "confession" from the pilot contained in response to the letter of investigation.

Some practitioners in this field advise that the letter of investigation should be ignored. Generally, I do not give such advice, but choose to respond to the letter of investigation on behalf of a particular client. The response must be carefully worded so that, considering the circumstances and what the FAA already knows, important pieces of evidence otherwise missing are not provided in that response. I have always felt that it is better to generally respond and treat the investigator with the courtesy of an answer, but let him know at the outset that he is dealing with counsel and that he will not be able to gain evidence that can be used against counsel's client.

Once the letter of investigation is answered, the FSDO inspector finishes his investigation within the FSDO by determining whether the violation appears to be one for which *certificate action*—suspension or revocation of the owner-pilot's particular certificate—is appropriate or called for in the enforcement and compliance policy manual. The investigator then ships his investigative file to the regional counsel at the FAA regional office governing the geographic area in which the alleged violation occurred, as well as the geographic area in which the FSDO is located. At that point, the FSDO inspector will send a form letter to the original recipient of the letter of investigation, telling him that the matter has been forwarded to regional counsel and that further contact will come from the region.

Once the letter of investigation is received, meaningful discussion concerning the allegations and the flight in question should occur only between the certificate holder and his counsel. In my opinion, many cases are lost because pilots and other certificate holders respond to the letter of investigation in a manner that not only helps the FAA's prosecution of its case, but often cements the eventual success of it.

Immediately upon receiving a letter of investigation, a pilot should seek the services of a competent aviation attorney and should not, himself, attempt to respond to the letter.

I have seen in my practice over the years many cases that might have been successfully defended, or at least the certificate holder might have been put in a far better bargaining position for eventual settlement, but because the cases were so compromised initially by the certificate holder's response to the letter of investigation, future defense approaches became severely limited.

A confession made in a response to a letter of investigation cannot be suppressed as can one procured from a criminal defendant who is not apprised of her rights. Be careful and tight-lipped.

NOTICE OF PROPOSED CERTIFICATE ACTION

When regional counsel receives the investigative file from the FSDO that initiated the process, the case will be assigned to an attorney on the regional counsel staff to conduct further investigation and propose a suitable penalty if it appears that the FAA has a reasonable chance of proving the alleged violation.

Assuming that the staff attorney feels that a provable case exists, the next contact that the certificate holder will receive is a very formal document known as a *notice of proposed certificate action*. This notice will, in all likelihood, come to the recipient by certified mail, and will look far more official than did the letter of investigation. The notice of proposed certificate action will set forth in numbered paragraphs the factual allegations that form the basis of the alleged violation. It will close with a statement that the FAA has determined that the appropriate penalty is a *suspension* (for a specified period of time) or *revocation* (permanent) of the recipient's certificate. A sample notice of proposed certificate action appears as Figure 6-2.

The notice of proposed certificate action cannot be ignored. If an appropriate response—to be discussed in a few paragraphs—is not forthcoming within the specified period of time (usually 20 days), this notice of proposed certificate action becomes final, with no further action necessary, and the then proposed suspension or revocation becomes a reality. Therefore, even if the certificate holder chose to ignore the initial letter of investigation, she had better not ignore the notice of proposed certificate action.

April 5, 19-- 88GL140000

CERTIFIED MAIL - RETURN RECEIPT REQUESTED

Captain Jonathan V. Doaks
25 Duval Drive
East Miami, Illinois

NOTICE OF PROPOSED CERTIFICATE ACTION

Take notice that upon consideration of the report of investigation, including a statement of February 8, 19--, made on your behalf by Mr. D. R. Roe, Senior Vice President, Flight Operations, Jones Airlines, it appears that you violated the Federal Aviation Regulations by reason of the following circumstances:

1. You are now, and at all times mentioned herein were, the holder of Airline Transport Pilot Certificate No. 1000000.

2. On or about January 15, 19--, you acted as pilot in command of a Jones Airlines Boeing 720 aircraft operating on instrument flight rules in air transportation as scheduled Flight 13 from O'Hare International Airport, Chicago, Illinois, to Willow Run Airport, Ypsilanti, Michigan.

3. During the above-described flight, Flight 13 was instructed by air traffic control (ATC) to maintain an altitude of 8,000 feet.

4. Notwithstanding said instruction, Flight 13 descended to an altitude of about 7,400 feet, 600 feet below the assigned altitude.

5. Your operation of Flight 13, in the manner and under the circumstances described above, was careless so as to endanger the life and property of another.

By reason of the foregoing facts and circumstances, you violated the following Federal Aviation Regulations:

1. Section 91.75(b), in that, in an area in which air traffic control is exercised, you operated an aircraft contrary to an ATC instruction without obtaining an amended instruction; and

2. Section 91.9, in that you operated an aircraft in a careless or reckless manner so as to endanger the life or property of another.

Please take notice that by reason of the foregoing facts and circumstances and pursuant to the authority vested in the Administrator by Section 609 of the Federal Aviation Act of 1958, as amended, we propose to suspend your airline transport pilot certificate for a period of 30 days.

Unless we receive, in writing, your choice of the alternatives provided and set forth on the enclosed information form, on or before ____date____, an order of suspension will be issued as proposed.

Assistant Chief Counsel

By: ______________________
Attorney

Enclosures

Fig. 6-2. Sample Notice of Proposed Certificate Action (Federal Aviation Act).

Options in Response to Notice of Proposed Certificate Action

The notice will be accompanied by a form that sets forth the options available in response. (See Figure 6-3.) The recipient can simply ignore the notice of proposed action and allow it to become final, and then file an appeal to the National Transportation Safety Board (NTSB). This is seldom a good approach, for reasons about to be discussed.

The first option shown in Figure 6-3 is to immediately surrender one's certificate. This action waives all rights of appeal to the NTSB—and is seldom preferable. About the only time it is used by knowledgeable practitioners is in a situation where both the client and the attorney realize that the FAA has merely scratched the surface of what might be much more serious or more numerous violations. It might well behoove the violator at that point to fold his tent and take whatever proposed action has been suggested, because if the investigation continues, further or more serious violations might be uncovered that could multiply his problems. Absent this rather rare occurrence, an immediate surrender of the certificate, and the accompanying waiver of all appeal rights, is rarely in order.

The second alternative that a certificate holder has when presented with a notice of proposed certificate action is to request that an order be issued to make final the certificate action that has been proposed, and then appeal that order to the National Transportation Safety Board.

When a certificate holder elects this option, she forgoes any further opportunity to negotiate and work toward a settlement of the controversy with the FAA attorneys, and proceeds directly to the appeals step subsequently discussed in this chapter. Almost the only time this course of action is appropriate is when long and protracted discussions and negotiations have already occurred with the FAA prior to the issuance of the proposed notice of certificate action. If the respondent and her attorney feel that any further such attempts to reach a compromise would be fruitless, and come to a reasoned belief that the certificate action is inevitable, sometimes the best approach is to request the FAA to proceed to issue the final order, so that the certificate holder may then directly appeal. Taking this step does not result in any waiver of appeal rights, as does the immediate surrender of the certificate under option 1 in Figure 6-3.

The third option available in responding to a notice of proposed certificate action is to answer the charges in writing. Again, seldom is this a good approach. Many of the same concerns surrounding the answer to the letter of investigation must be kept in mind if a certificate holder chooses to answer a notice of proposed certificate action in writing. Whatever that answer is will become a part of the file. If it is incriminating, the respondent can rest assured that his own statements will be used against him. Before any such charge is answered in writing, it is absolutely imperative that the certificate holder obtain the services of an attorney and allow the attorney to compose such an answer if one is to be used. Rarely do practitioners afford themselves this option. That is simply because virtually anything that needs to be said can be expressed at an informal conference, subsequently discussed in this chapter.

INFORMATION WITH RESPECT TO NOTICE OF PROPOSED CERTIFICATE ACTION

You may elect to proceed in one of the five ways set forth below. You should use the enclosed form for replying to the notice of proposed certificate action to indicate how you elect to proceed. You may proceed as follows:

1. Surrender your certificate on or before the above date. In this event, the order proposed in the notice will be issued at once, effective the date your certificate is surrendered or mailed to the office listed below. By surrendering your certificate, you will waive your right to appeal to the National Transportation Safety Board, as described in 2 below.

2. Indicate your desire to have an order issued as proposed in the notice of proposed certificate action so that you may appeal to the National Transportation Safety Board (NTSB), as provided in Section 609 of the Federal Aviation Act of 1958, as amended (49 U.S.C. 1429) and the NTSB's Rules of Practice (49 C.F.R. Part 821). This may be done by checking item no. 2 on the enclosed reply form or by not responding to the notice. You may proceed in accordance with 3, 4, or 5, below, without waiving your right to appeal to the NTSB.

If you appeal to the NTSB, an administrative law judge will be appointed to decide the case. The judge may hold a formal hearing at which the FAA will present witnesses and other evidence, and you will have the opportunity to present witnesses and other evidence. The FAA will have the burden of proof. An appeal from the judge's decision to the full Board is available, and from there to the U.S. courts of appeal.

3. Answer the charges in writing. With such answer, you may furnish such additional information, including statements by you or your representative or others, or other documentary evidence as you may wish to have considered. This will not affect your right to appeal to the NTSB.

4. Request that you and/or your representative be accorded a conference with an FAA attorney at the regional office of the FAA or at ______________________________. At this conference, you may state why the proposed action should not be taken and you may present evidence and information on your behalf. This will not affect your right to appeal to the NTSB.

5. If you have filed an Aviation Safety Report with the National Aeronautics and Space Administration (NASA) concerning the incident set forth in the attached notice of proposed certificate action, you may be entitled to waiver of any penalty.

You will only be entitled to waiver if it is found --

a. That this violation was inadvertent and not deliberate;

b. That this violation did not involve a criminal offense, or accident, or disclose a lack of competence or qualification of the holder of a certificate; and

c. You have not paid a civil penalty pursuant to Section 901 of the Federal Aviation Act or been found in any prior FAA enforcement action to have committed a violation of the Federal Aviation Act, or any regulation under the Federal Aviation Act, for a period of 5 years prior to the date of the occurrences.

d. You prove that within 10 days after the violation, you completed and delivered or mailed a written report of the incident or occurrence to NASA under the Aviation Safety Report Program.

In the event that you prove your entitlement to this waiver of penalty, an order will be issued finding you in violation but imposing no certificate suspension. Your claim of entitlement to waiver of penalty shall constitute your agreement that this order may be issued without further notice. You will, however, have the right to appeal the order to the National Transportation Safety Board (NTSB) pursuant to Section 609 of the Federal Aviation Act.

Fig. 6-3. Sample information sheet to accompany Notice of Proposed Certificate Action.

Following the issuance of an order, you will have the right to appeal such order to the NTSB under the provisions of Section 609 of the Federal Aviation Act.

Address all communications in this matter to:

Assistant Chief Counsel
Federal Aviation Administration
Address

If the certificate holder is an individual:

PRIVACY ACT NOTICE

This notice is provided in accordance with Section (e)(3) of the Privacy Act, 5 U.S.C. Section 552a(e)(3), and concerns the information requested in the letter or form with which this Notice is enclosed.

A. Authority. This information is solicited pursuant to the Federal Aviation Act of 1958, 49 U.S.C. Section 1301, et seq., and regulations issued thereunder codified in Part 13 of Title 14 of the Code of Federal Regulations. Submission of the telephone number is voluntary. The request for information is intended to provide you with an opportunity to participate in the investigation.

B. Principal purpose. The requested information is intended to assist us in contacting you regarding this enforcement case.

C. Routine uses. Records from this system of records may be disclosed in accordance with the routine uses as they appear in System of Records No. DOT/FAA 847 as published from time to time in the Federal Register.

D. Effect of failure to respond. If you do not provide the requested information, there may be delay in contacting you regarding this enforcement case.

Fig. 6-3. *Continued.*

Most practicing attorneys prefer to deal with the opposition face to face, rather than through the medium of written answers. There is certainly a place for written answers, and as far as these proceedings go, that place is at the stage of a formal appeal to the National Transportation Safety Board. But in response to a notice of proposed certificate action, written answers are not often submitted. One has a difficult time retracting any answer that has been given in writing. Therefore, if either the certificate holder or his attorney learns of further facts that would tend to contradict the written answer, he is immediately put at a disadvantage in that, in essence, he has to change his story.

The fourth option, and the one most frequently used, is the ability to request what is known as an *informal conference* in order to discuss the issues with the FAA staff attorney who will be prosecuting the case. This conference will usually occur at an FSDO location convenient for the certificate holder. The FAA staff attorneys tend to ride circuit throughout their region, and quite often the informal conference will be months in coming. Because of constraints within the federal budget, there is a movement afoot to conduct informal conferences by telephone, to save the FAA the expense of sending its attorneys all over their respective regions to conduct informals in person. Many members of the legal profession who defend these types of cases, I included, feel that some courts may hold that telephone conferences will not meet the FAA's legal requirement to give the certificate holder an "opportunity to be heard" before a certificate is formally suspended.

Even if the move toward telephone informals takes hold, and survives the inevitable court challenges, I would still prefer to hold an informal conference in most cases. There are two primary reasons to have an informal conference, even if it must be by telephone.

The first reason is that, if the certificate holder requests it, pursuant to the federal Freedom of Information Act (FOIA) a good portion of the FAA's file must be turned over to the attorney whom the certificate holder has hopefully retained by this time. Recommendations contained in interoffice memos within the FAA and certain portions of the investigative file are not subject to release, but by and large the greater portion of the file will be disclosed to the certificate holder.

At this stage, the certificate holder, also known as the *respondent*, will have a good idea of the strength of the FAA's case against him. This will enable the defense attorney, in consultation with her client, to determine an approach, tactics, strategy, and philosophy of defense for the remainder of the proceeding.

The second reason for requesting the informal conference is that the rules require that, once requested, the informal conference must be provided by the FAA. If the FAA fails to so provide this opportunity, there is a good legal argument that a jurisdictional requirement has not been met, and that any further attempts by the FAA to seek a certificate suspension or revocation for this particular alleged violation should not be allowed to go forward.

As with other such requirements, the FAA seldom misses this one. But in the event that it does, the certificate holder has an extremely good defense on jurisdictional grounds that will, in all likelihood, end the matter and result in a complete victory for the respondent. Most practitioners who are experienced in this area take advantage of the opportunity to perhaps catch the adversary off guard by requesting an informal conference.

If the certificate holder has not retained competent counsel at this point, he certainly should. It is never wise to go into an informal conference unrepresented, because at this point a layperson is dealing with an FAA attorney. Most of the FAA attorneys are extremely competent, practice in a very narrow and limited area, and become experienced and savvy in short order. For a pilot, mechanic, or other certificate holder to face this formidable opposition unrepresented borders on the foolhardy, unless the respondent is already resigned to the fact that he is willing to go along with the proposed certificate action and is simply going through the motions at this point.

As with other contacts with the FAA, the certificate holder needs to be extremely guarded in the comments that he makes at the informal conference. A slip of the tongue has hung more than one person in this situation, and invariably will in the future.

Overall, the option most often exercised in response to the notice of proposed certificate action is the conduct of the informal conference. This option seldom has a serious downside risk, and therefore should be carefully considered, and usually exercised.

The fifth option stated on the information sheet that accompanies the notice of proposed certificate action, which appears as Figure 6-3, concerns the *immunities* available to a certificate holder who has filed an aviation safety report with the National Aeronautics and Space Administration (NASA) concerning the incident set forth in the notice. A lengthy discussion of the aviation safety reporting system occurs later in this chapter, so our review of option 5 in Figure 6-3 shall be deferred until that time.

Figure 6-4 is a sample of another document that will accompany a notice of proposed certificate action. On this form, the certificate holder replies to the FAA as to which of the five alternative courses of action she elects. Seldom is more than one of the five options appropriate, except when an aviation safety report has been filed, whereupon it is generally appropriate to check both options 4 and 5. In that case, an informal conference can be held, at which time the effect of the filing of the aviation safety report can be discussed with the FAA attorney who will be prosecuting the case, assuming the FAA has not already seen fit to accept the report and deal with it accordingly.

THE STALE COMPLAINT RULE

Under the section of the rules of practice that governs appeals to the National Transportation Safety Board, the recipient of a notice of proposed certificate action must be put on notice of the complaint against him within 6 months of the alleged violation—the stale complaint rule. Therefore, it is extremely important that the FAA send out its notice of proposed certificate action within this time limitation.

Date: ____________________

To: Assistant Chief Counsel
Address

Subject: Notice of Proposed Certificate Action

In reply to your notice of proposed certificate action and the accompanying information sheet, I elect to proceed as indicated below:

1. /__/ I hereby transmit my certificate with the understanding that an order will be issued as proposed, effective the date of mailing of this reply, and with the understanding that I waive my right to appeal the order to the National Transportation Safety Board.

2. /__/ I request that the order be issued so that I may appeal directly to the National Transportation Safety Board.

3. /__/ I hereby submit my answer to your notice and request that my answer and any information attached thereto be considered in connection with the allegations set forth in your notice.

4. /__/ I hereby request to discuss this matter at an informal conference with an attorney from your office at ____________________
____.

5. /__/ I hereby claim entitlement to waiver of penalty under the Aviation Safety Report Program and enclose evidence that a timely report was filed with NASA.

Certificate holder:	My attorney/representative:
Signature: ____________	Name: ____________
Address: ____________	Address: ____________
____________	____________
Telephone: ____________	Telephone: ____________

Fig. 6-4. Sample certificate holder reply.

Be watchful, for occasionally a staff attorney within the regional counsel's office will err, or an FSDO inspector will take too long in his investigation and exceed this time limit. If more than 6 months has elapsed, the FAA still has an option to pursue a civil penalty against the alleged violator, but unless it can show good cause for the delay, the passage of 6 months will prevent a certificate action. One should not assume that this 6-month period of time is concrete, since the rule does allow the FAA to proceed in a certificate action if it can show good cause for the delay. Such good cause can be sometimes shown if the investigative period was necessarily long and the alleged violation complex. Nevertheless, the stale complaint rule exists, and should always be used as a potential defense if the 6-month time limit is exceeded.

NOTICE OF CERTIFICATE ACTION

Throughout the procedure just described, the FAA has not yet issued any final order to actually suspend the certificate in question—unless the certificate holder elected option 1 or 2, as set forth in Figure 6-4. Because neither of those possible responses is often used, the actual *notice of certificate action* will not be issued until the conclusion of the informal conference or after consideration by the FAA of whatever further written material the certificate holder and her attorney submit, in the event they take that approach.

The notice of certificate action will be an official order of the FAA either suspending or revoking the certificate in question. A sample of an order of suspension is set forth in Figure 6-5. You will note that it is composed in a format very similar to the notice of proposed certificate action. Perhaps during the informal conference or through other negotiations or information that the FAA might have at its disposal, a change will be made in some substantive allegation. In that event, the order of suspension or revocation could possibly differ somewhat from the notice of proposed certificate action. This is a rare occurrence, however, and generally the order of suspension will closely parrot the wording of the notice of proposed certificate action.

The order of suspension or revocation, referred to here as the *notice of certificate action,* is a very important document. Obviously, it is important because it is the official manner in which the FAA formally suspends or revokes the certificate in question. But just as crucially, it sets forth the factual allegations upon which the proposed suspension or revocation is based and, further, cites the particular regulations that the certificate holder is accused of violating. When an appeal to the NTSB is taken from this order, the order will be the actual complaint of the FAA in these quasi-judicial proceedings. A very careful review is in order of both the factual allegations and the regulations that are claimed to be violated. If there is a factual allegation which can be disputed, or in which the FAA has made an error, so much the better for the certificate holder.

The order of suspension, in a manner similar to the notice of proposed certificate action, cannot be ignored unless the respondent desires to allow it to become effective and wishes to suffer the ordered suspension or revocation.

May 16, 19--   88GL140000

CERTIFIED MAIL - RETURN RECEIPT REQUESTED

Captain Jonathan V. Doaks
25 Duval Drive
East Miami, Illinois

ORDER OF SUSPENSION

On April 5, 19--, you were advised by mail through a notice of proposed certificate action of the reasons why we proposed to suspend your airline transport pilot certificate, No. 1000000, for a period of 30 days.

After a consideration of all the evidence presently a part of this proceeding, including the information you presented at the informal conference held in the Office of the Assistant Chief Counsel, Kansas City, Missouri, on May 9, 19--, it has been determined that you violated the Federal Aviation Regulations because of the following circumstances:

1. You are now, and at all times mentioned herein were, the holder of Airline Transport Pilot Certificate No. 1000000.

2. On or about January 15, 19--, you acted as pilot in command of a Jones Airlines Boeing 720 aircraft operating on instrument flight rules in air transportation as scheduled Flight 13 from O'Hare International Airport, Chicago, Illinois, to Willow Run Airport, Ypsilanti, Michigan.

3. During the above-described flight, Flight 13 was instructed by air traffic control (ATC) to maintain an altitude of 8,000 feet.

4. Notwithstanding said instruction, Flight 13 descended to an altitude of 7,400 feet, 600 feet below the assigned altitude.

5. Your operation of Flight 13, in the manner and under the circumstances described above, was careless so as to endanger the life and property of another.

By reason of the foregoing facts and circumstances, you violated the following Federal Aviation Regulations:

1. Section 91.75(b), in that, in an area in which air traffic control is exercised, you operated an aircraft contrary to an ATC instruction without obtaining an amended instruction; and

2. Section 91.9, in that you operated an aircraft in a careless or reckless manner so as to endanger the lives and property of another.

By reason of the foregoing, the Administrator has determined that safety in air commerce or air transportation and the public interest require the suspension of your airline transport pilot certificate.

NOW, THEREFORE, IT IS ORDERED, pursuant to the authority vested in the Administrator by Section 609 of the Federal Aviation Act of 1958, as amended, that your airline transport pilot certificate, No. 1000000, be suspended, effective May 23, 19--, said suspension to continue in force for a period of 30 days thereafter. In the event you fail to surrender your certificate to the Office of the Assistant Chief Counsel, Federal Aviation Administration, Address, on or before May 23, 19--, said suspension shall continue in effect until 30 days subsequent to the actual surrender thereof.

Assistant Chief Counsel

By:__________________________
Trial Attorney

APPEAL

You may appeal from this order within 20 days from the date it is served by filing a Notice of Appeal with the Office of Administrative Law Judges, National Transportation Safety Board, Room 822, 800 Independence Ave., S.W., Washington, D.C. 20594.

Part 821 of the Board's Rules of Practice (49 C.F.R. Part 821) applies to such an appeal. An original and three (3) copies of your appeal must be filed with the National Transportation Safety Board (NTSB). In the event you appeal, a copy of your notice must also be furnished to the Office of Chief Counsel/Office of Assistant Chief Counsel at the address noted in the foregoing Order.

Fig. 6-5. Sample Order of Suspension (Federal Aviation Act).

The filing of a timely appeal will stay the effectiveness of this Order during the pendency of that appeal before the NTSB. If you appeal to the NTSB, a copy of this Order will be filed with the NTSB and will serve as the Administrator's complaint in this proceeding.

CERTIFICATE OF SERVICE

I hereby certify that this Order has been mailed this date by certified mail, addressed to:

Legal Clerk   Date

Fig. 6-5. *Continued.*

If the certificate holder has disputed the case thus far, it is a rare occurrence that he will simply want to go along at this point. One large exception to the immediately preceding statement occurs when, through negotiation at, before, or subsequent to the formal conference, an agreement is made between the FAA attorney and the certificate holder to compromise the case in some manner.

An order of suspension or revocation will then be issued, and must be reviewed carefully to make sure that it complies with the agreement made between the certificate holder and the FAA attorney. Remember, the only way the FAA can suspend or revoke any certificate is through this official order. Even though the case has possibly been settled or compromised along agreeable lines, if the agreement involves any suspension or revocation of the certificate in question, a notice of certificate action must still be issued in order for that settlement to be carried into effect.

EMERGENCY ENFORCEMENT ACTIONS

Up until this point, we have been discussing the enforcement procedure used to suspend, or sometimes revoke, a certificate in which the certificate remains valid and in force during the enforcement process. If the FAA chooses to utilize its authority to declare an enforcement matter to be an emergency, that is not the case.

Emergency proceedings do not occur very often, but they are a virtual certainty for certain alleged violations. In my experience, the FAA will assert an emergency when it catches illegal Part 135 operations, which means that it finds someone running charter flights without the required Part 135 certificate.

Other areas where emergency cases will be brought surround cheating and other forms of dishonesty. I have represented pilots who were the subject of emergency revocations of their pilot certificates when they were found with "crib" sheets, or other methods for cheating on subsequent written examinations. I dealt once with a client who was accused of padding the time in his logbook, when he presented himself for a flight test. The FAA inspector sensed that something was wrong with the logbook as he was examining it. Cheat, pad, or run illegal charters, and you virtually guarantee yourself that you'll learn, firsthand, how the emergency enforcement procedures work.

The FAA takes the attitude, in most cases, that if the alleged violation warrants revocation of the certificate, the case ought to be an emergency. This philosophy holds that, since revocation is such a severe remedy, reserved for the worst offenders, these people ought not to be exercising their certificates during the protracted period of time, often more than a year, that it takes a normal case to wind its way through the system.

The certificate holder is grounded, put out of work, or whatever, during the emergency enforcement procedure. Because of this draconian result, all the procedural timetables are greatly compressed. If the certificate holder is innocent, the system wants to let him or her get the privileges of the certificate restored as soon as possible.

If the FAA decides to go with an emergency case, there will not be any notice of proposed certificate action. Rather, the certificate holder will receive an emergency order, usually of revocation, which receipt starts a very fast clock ticking. The recipient of the order has only 10 days in which to file the notice of appeal with the NTSB.

After the notice of appeal is filed, the FAA must file its complaint within 3 days of the date that the notice of appeal was received by the NTSB. Within another 5 days, the certificate holder must file an answer to that complaint. Then, the hearing before the NTSB law judge must be noticed to the parties, and actually held within 7 more days after the notice goes out. The law judge is required to issue the initial decision orally at the end of the hearing.

If either side wants to appeal the initial decision, the notice of appeal of that result must be filed with the NTSB within 2 days after the initial decision. The time schedules for filing the various briefs and arguments to the full board are likewise very tight.

That way, fairness is accorded its due. If the certificate holder prevails, the inconveniences of being denied the privileges of the certificate are kept to a minimum. If the FAA wins, the facts of the misconduct were so severe that, according to the philosophy behind emergency cases, the public deserved the protection of the certificate holder's inability to use the certificate during the pendency of the process.

The best way to view emergency cases is to avoid being caught up in one. Remember that this severe process is reserved for the folks that the FAA deems to be the real bad guys, and it isn't used for the everyday violation that results from neglect or inadvertence. Don't cheat, don't run illegal charters, don't fly instruments without an instrument rating. If you obey those commands, and ones of similar import, you'll only read about emergency cases.

THE AVIATION SAFETY REPORTING SYSTEM

For several years now, NASA has had a program to identify potential hazards to aviation, and a mandate to issue periodic reports that are to be used by everyone within the aviation community: the FAA, NTSB, and NASA, as well as all the pilots, mechanics, fixed-base operators, airlines, and any other concerned person or organization.

In order to receive frank and honest reports about potential dangers within the system, it was decided, when the aviation safety reporting system (ASRS) began in its most recent form, that input from users was absolutely vital for the success of this endeavor.

Several decades ago, a similar program was tried, and reporting forms were sent to pilots and other certificate holders for use in reporting hazards and other problems encountered. However, it was often a catch-22 under this former program, because when someone reported a problem that involved some action or inaction that could possibly be a regulatory violation, he had no guarantee that his report would not be used against him and, quite possibly, form the basis for an FAA enforcement proceeding.

To say the least, the first program was not particularly successful, as people quickly learned that silence was the best course of action. When the process was revised, and the current ASRS devised, NASA and the FAA realized that, in order for reports to be submitted with frankness and candor, some form of immunity must accompany such voluntary action on the part of certificate holders. For that reason, the current system involves a limited degree of immunity and can be used openly and honestly by a user of the national aviation system without concern that filing such a report will come back to haunt the person who took the time to complete it and further the goals of aviation safety.

Figure 6-6 shows the current aviation safety report form. The greatest guarantee given to the reporter by this program is the fact that the reporting form remains totally anonymous.

At the very top of the front of the form is a strip whereon the reporter puts his name and mailing address. When this form then arrives at NASA, the form is de-identified by means of cutting the identification strip from the top, time-stamping it to indicate the date of receipt by NASA, and returning it to the sender. Therefore, the form that remains in NASA's hands has no information on it that can identify either the sender or the particular aircraft involved in the incident.

Completing an aviation safety report and filing it with NASA will give the reporter immunity from certificate action or civil penalty under certain conditions, the first of which is most important: *The ASRS report must be completed and sent to NASA within 10 days of the occurrence of the incident being reported.* This time limit was conceived to serve a useful purpose. Neither the FAA nor NASA wanted the report to be used only as a shield for violation cases, and both desired that the report be sent in the interests of aviation safety, before the reporter was aware of any violation proceeding that might be forthcoming against her.

In today's world, for certain violations, certificate holders might well know immediately or shortly after the occurrence that the FAA will be prosecuting the case, but for a good number of such cases, the certificate holder does not know what the eventual outcome of the incident will be. Therefore, the purpose of the aviation safety report is achieved when all users are encouraged to file the report promptly, and not simply sit back and wait to see if the FAA will be taking action against them before taking the trouble to complete the report and send it in.

Therefore, anytime a certificate holder feels that something has occurred that might form the basis for an enforcement action against him, in the nature of either a certificate action or a civil penalty, he should immediately fill out an aviation

IDENTIFICATION STRIP: *Please fill in all blanks.* ***NO RECORD WILL BE KEPT OF YOUR IDENTITY.***
This section will be returned to you promptly.

TELEPHONE NUMBERS where we may reach you for further details of this occurrence:

(HOME) Area_____ No._____ - ________ Hours_____
(WORK) Area_____ No._____ - ________ Hours_____

(SPACE RESERVED FOR ASRS DATE/TIME STAMP)

NAME ____________________
ADDRESS ____________________

TYPE OF EVENT/SITUATION ____________________
DATE OF OCCURRENCE ____________________
LOCAL TIME (24 hr. clock) ____________________

Except for reports of aircraft accidents and criminal activities – which are not included in the ASRS and should not be submitted to NASA – all identities contained in this report will be removed to assure complete reporter anonymity.

PLEASE FILL IN APPROPRIATE SPACES AND CHECK ALL ITEMS WHICH APPLY TO THIS EVENT OR SITUATION.

REPORTER'S ROLE DURING OCCURRENCE
(pilot-flying, radar controller, cabin crew, maintenance, etc.) ____________________

REPORTER	FLYING TIME	CERTIFICATES/RATINGS		ATC EXPERIENCE
○ captain/pilot	total _____ hrs.	○ student	○ private	○ FPL ○ developmental
○ first officer	last 90 days _____ hrs.	○ commercial	○ ATP	radar _____ yrs.
○ other crewmember	in acft type _____ hrs.	○ instrument	○ CFI	non-radar _____ yrs.
○ controller		○ multiengine	○ F/E	supervisory _____ yrs.
○ ________		○ ________		military _____ yrs.

DESCRIBE ONE AIRCRAFT IN THIS SECTION (PILOTS DESCRIBE YOUR OWN) AND ADDITIONAL AIRCRAFT IN THE "DESCRIBE EVENT/SITUATION" SECTION:

AIRFRAME/ENGINES			OPERATOR		PURPOSE OF FLIGHT	FLIGHT PLAN
○ low fixed wing	○ ultralight	○ reciprocating	○ scheduled carrier		○ passenger	○ VFR ○ IFR
○ high fixed wing	○ wide body	○ turboprop	○ supplemental carrier		○ cargo	○ SVFR ○ none
○ rotary wing	○ small complex	○ turbojet	○ FBO/flying school		○ business	**NAVIGATION IN USE**
○ advanced/automated cockpit (e.g., CRT's, FMS, etc.)			○ commuter	○ air taxi		
○ ________			○ corporate	○ charter	○ training	________
crew size _____	pax seats _____		○ government	○ private	○ pleasure	________
gross weight _____	no. of engines _____		○ military (_____)		○ _____	________
			○ ________		○ _____	________

FIRST FOLD HERE

AIRSPACE/LOCALE			ATC/ADVISORY SERVICE		FLIGHT CONDITIONS		LIGHT AND VISIBILITY	
○ uncontrolled	○ ATA	○ PCA	○ ground	○ approach	○ VMC	○ IMC	○ daylight	○ dawn
○ control zone	○ TRSA	○ TCA	○ local	○ departure	○ mixed	○ marginal	○ dusk	○ night
○ special use airspace	○ ARSA	○ unknown	○ center	○ FSS	○ t'storm	○ rain	ceiling _____ feet	
○ airway/route	○ MTR	○ _____	○ UNICOM	○ CTAF	○ turbulence	○ fog	visibility _____ miles	
ALTITUDE _____	○ MSL (or) ○ AGL		Name of ATC Facility _____		○ windshear	○ snow	RVR _____ feet	
NEAREST CITY _____	STATE ___		________		○ ice	○ _____		

SPECIFY LOCATION BY REFERENCE TO AN AIRPORT, NAVAID, OR OTHER FIX (distance, bearing, etc.): ____________________

AIRCRAFT FLIGHT PHASES AT TIME OF OCCURRENCE (preflight, takeoff, cruise, hover, etc.) ____________________

IF A CONFLICT: Evasive action? ○ yes ○ no ○ no time ○ unknown. Estimated miss in feet ______ vert'l ______ horiz'l.

DESCRIBE EVENT/SITUATION

Keeping in mind the topics shown below, discuss those which you feel are relevant and anything else you think is important. Include what you believe really caused the problem, and what can be done to prevent a recurrence, or correct the situation. (CONTINUE ON THE OTHER SIDE AND USE ADDITIONAL PAPER IF NEEDED).

CHAIN OF EVENTS		HUMAN PERFORMANCE CONSIDERATIONS	
– How the problem arose	– How it was discovered	– Perceptions, judgements, decisions	– Actions or inactions
– Contributing factors	– Corrective actions	– Factors affecting the quality of human performance	

NASA ARC 277(Rev. Oct 84) PREVIOUS EDITIONS ARE OBSOLETE

Fig. 6-6. Current aviation safety report form.

National Aeronautics and
Space Administration

Ames Research Center
Moffett Field, California 94035

Official Business
Penalty for Private Use $300

NO POSTAGE NECESSARY IF MAILED IN THE UNITED STATES

BUSINESS REPLY MAIL
FIRST CLASS PERMIT NO. 12028 WASHINGTON, D.C.

POSTAGE WILL BE PAID BY NASA

NASA Aviation Safety Reporting System
Post Office Box 189
Moffett Field, California 94035

FIRST CLASS
AVIATION SAFETY DATA —
DO NOT DELAY

NASA

NATIONAL AERONAUTICS AND SPACE ADMINISTRATION

NASA has established an Aviation Safety Reporting System to identify problems in the aviation system which require correction. The program of which this system is a part is described in detail in FAA Advisory Circular 00-46C. Your assistance in informing us about such problems is essential to the success of the program. Please fill out this postage free form as completely as possible, fold it and send it directly to us.

The information you provide on the identity strip will be used only if NASA determines that it is necessary to contact you for further information. THE IDENTITY STRIP WILL BE RETURNED DIRECTLY TO YOU. The return of the identity strip assures your anonymity.

AVIATION SAFETY REPORTING SYSTEM

Section 91.57 of the Federal Aviation Regulations (14 CFR 91.57) prohibits reports filed with NASA from being used for FAA enforcement purposes. This report will not be made available to the FAA for civil penalty or certificate actions for violations of the Federal Air Regulations. Your identity strip, stamped by NASA, is proof that you have submitted a report to the Aviation Safety Reporting System. We can only return the strip to you, however, if you have provided a mailing address. Equally important, we can often obtain additional useful information if our safety analysts can talk with you directly by telephone. For this reason, we have requested telephone numbers where we may reach you. Thank you for your assistance.

NOTE: AIRCRAFT ACCIDENTS SHOULD NOT BE REPORTED ON THIS FORM. SUCH REPORTS SHOULD BE FILED WITH THE NATIONAL TRANSPORTATION SAFETY BOARD AS REQUIRED BY 49CFR830.

15. NARRATIVE DESCRIPTION (continued): ***(Use additional sheets if necessary)***

SECOND FOLD HERE

SECOND FOLD HERE

Fold as indicated, fasten with staple or tape, and mail. Thank you for your cooperation.

Fig. 6-6. *Continued.*

safety report and send it to NASA. I advise pilots, mechanics, authorized inspectors, fixed-base operators, and all other certificate holders to keep a supply of these forms, which can be obtained from the local FSDO office, on hand for immediate use whenever any such incident occurs.

When the safety report is completed, the reporter has two options to guarantee that the report is filed within 10 days. He can wait until he gets his identification strip back from NASA, which will be time-stamped with the time and date of receipt. He can then use that strip to prove, should he need to later on, that the form was sent within 10 days. But an argument can always arise as to the necessary time for mailing and transmittal by the postal service of any piece of mail. If the identification strip shows that the report was received on the eleventh day after the incident, most FAA attorneys and administrative law judges will assume it was mailed within the 10-day time period required. Much beyond that, though, and the certificate holder might have a serious argument on his hands as to whether he was diligent in complying with the absolute requirement that the report be made within 10 days.

The better course of action is to mail the report from a post office and request and receive a certificate of mailing, or to mail the report to NASA via certified mail, return receipt requested. By doing this, a certificate holder would have an even better degree of protection, being able to prove that he complied with the 10-day time requirement.

If the reporter feels that the incident is such that enforcement action is a possibility, it is well worth the minimal expense to take the report, once completed, to a post office and either receive a certificate of mailing or mail the document by certified mail with return receipt requested. A bit of thoroughness and diligence may pay off in the end, if any unexplained delay in the mailing process occurs and the identification is returned with a time-stamped date on it that could lead an objective observer to believe that the report was not sent within 10 days.

The filing of an aviation safety report will entitle the reporter to a waiver of any penalty for the incident that is described in the report, under certain limited conditions.

The first such condition is that the violation was inadvertent and not deliberate. If the FAA argues, and later proves, that the conduct was not inadvertent or that it was deliberate, the filing of an aviation safety report will not accomplish any immunity for the reporter. But again, the comments made within the report cannot be identified or retrieved by the FAA under any circumstances. So even if there is a question concerning inadvertent or deliberate conduct, it is still the better course of action to go ahead and file the report within the time limit.

Another *condition of entitlement* to a waiver of all penalties is that the violation did not involve a criminal offense or accident, or disclose any lack of competency or qualification of the holder of the certificate. In this regard, the first part of this condition is obvious. If a criminal offense is part of the violation, forget immunity. The ASRS is not going to protect drug smugglers, aircraft thieves, or other such persons. Next, if an accident occurs, the immunity is not present. Remember that the immunity offered through the aviation safety reporting system is very qualified; accidents are sufficiently serious business that the program will not provide immunity to someone who is involved in an accident by reason of or through violation of the FAR.

Lastly, as a part of this condition, if the violation discloses a basic lack of competence or qualification on the part of the holder of the certificate, immunity will also not be available. Situations that involve this scenario might be flight under instrument conditions without an appropriate instrument rating or any flight or ground operation that plainly discloses that the violator simply does not know what he is doing, and is not competent or qualified to attempt the operation in question.

The same comment can be made about mechanics, those with inspection authorization, and other certificate holders who are not actually flying aircraft. But in order for immunity to be denied for this reason, the violation must disclose a very basic and plain incompetence or lack of qualification; otherwise, immunity will be available.

If a previous violation results in an admission or adjudication of a violation of either the Federal Aviation Act or any regulation within the preceding 5 years, the ASRS will not grant immunity to the reporter for the filing of a report.

As to this last condition under which immunity is not available, it is important to realize that we are not talking about having filed a report within the last 5 years, but rather, either having paid a compromise civil penalty or actually been found in violation in the past 5 years. I personally know of many operators who frequently file reports, not only for safety reasons, but also to protect themselves from enforcement activity.

With the current enforcement attitude requiring very strict compliance with the regulations, there is no limit to the number of aviation safety reports one may file, or the frequency with which they may be sent. Do not be confused with this 5-year rule; it applies only to actual violations or compromise civil penalties paid within the preceding 5 years, not reports sent to NASA.

I carry the report forms in my flight bag, and have used several of them, even though I've never yet had a violation case filed against me. One day in 1990, I used two forms the same day. I was traveling from Columbus, Ohio up to Cleveland to visit a client, intending to land at Burke Lakefront Airport. Part of the flight up from the south involved transiting what was then the Cleveland Terminal Control Area (TCA), now called Class B airspace.

I called the controller as required, and got the typical VFR clearance through the TCA. When I was about 15 miles from Lakefront, the cabin speaker in my Comanche went dead as a doornail, and I could not hear a thing from the radios. I kept on going, and diverted to an uncontrolled field, landed, and called Cleveland Approach on the phone and explained what happened. Thankfully, the smaller airport wasn't far from Lakefront, so the client met me there. Even though I didn't note any adversary tone in the voice of the person to whom I spoke on the phone, I made a note to file a NASA report when I got back home.

The client happened to be an A&P mechanic. We bought a new speaker and installed it in the airplane after we were finished with our meeting, then I launched back for Columbus.

I was talking with Akron-Canton Approach as I was climbing southbound, since there was an ARSA (now Class C airspace) around Akron-Canton Airport. The controller asked me to once again say my type aircraft, and I responded that I was in a PA-24 Comanche. He then asked if it were a turbo Comanche. I responded in the negative, and asked him why he wondered if my airplane was turbocharged. His answer was that my Mode C encoder showed an altitude of 17,500 feet, and climbing. We both quickly agreed that the thing had gone haywire, and I was instructed to turn off the Mode C function. So far, not a good day.

The last part of the flight home entailed penetrating the Columbus ARSA (again, now Class C airspace) to get back to my home base at Ohio State University Airport. When I called Columbus Approach, I explained what was wrong with the transponder's encoder, and I was handled by approach control with no problem, and finally landed at OSU.

I had two possible problems with the FARs. One was operating in a TCA without maintaining radio contact; and the other resulted from being in two ARSAs without a functioning altitude encoder. So I spent 10 minutes and whipped off two NASA reports. I never heard a thing from the FAA about any of the day's misfortunes, but the avionics technician then at OSU Airport, affectionately nick-named "Radio Rick," got some more work.

I also started making a point of never leaving the ground without a headset. I learned to fly before headsets were worn in light planes. Although I had always worn them in helicopters, which routinely have the noise level of the engine room of a battleship, I wasn't used to using one in an airplane. I do now.

In summary, the ASRS serves a very worthwhile goal, being conceived in such a manner as to allow users of the system to file reports for hazards and dangers that they encounter, as well as for situations that simply suggest improved procedures. A very important side benefit, however, is the immunity from penalty offered to the reporter.

Remember, though, that the reporting system—although offering an immunity from penalties under the conditions discussed above—does not prevent the finding of an actual violation. The violation can be found by the FAA, and prosecuted, just the same as if the reporting form had not been filed. If all the conditions are met, the certificate holder will suffer no penalty; still, all the other detrimental effects of a violation will remain. The detrimental effects might include such things as increased insurance premiums, lack of opportunity to gain certain employment (if the employer is not willing to hire a person with a particular violation on his record), and similar stigma that might follow the violator for years.

Because the reporting system allows immunity only once every 5 years, the FAA will more than likely pursue a violation case to get a finding of violation on the record so that, should a further violation occur within the next 5 years, immunity would not be available.

Therefore, the ASRS should be viewed, not as a shield that cannot be penetrated, but simply as a protective device offering whatever immunity is available under it. It is not a license to violate the regulations, and violators will still be dealt with appropriately by the FAA, even though immunity from penalties might be present. Use the reporting system as it was designed—as an aid to aviation safety, not a program that has been ill defined and ill conceived to enable some people to violate the regulations with a cavalier attitude toward the consequences.

APPEAL TO THE NATIONAL TRANSPORTATION SAFETY BOARD

Once an order of suspension or revocation, also known as a notice of certificate action, has been issued to the certificate holder, the recipient has basically two choices. The first is to concede and send the certificate to the FAA attorney who issued the notice, and thereby suffer the ordered suspension or revocation. The other alternative, and the one frequently used, is to appeal the order to the NTSB.

The National Transportation Safety Board has several functions, and so far as they affect general aviation, the two most often thought of are the *investigatory function* and the *appellate function.* The NTSB investigates, through its field investigators, all fatal accidents, and many serious but nonfatal accidents as well. The other role of the NTSB that directly affects general aviation, as well as air carrier operators—and into which we will delve here—is the board's appellate function.

Title 49 of the *Code of Federal Regulations*, in Part 821, sets out the appellate jurisdiction of the NTSB, and subpart (D) of Part 821 deals with the rules applicable to NTSB appellate proceedings under section 609 of the Federal Aviation Act. It is section 609 of the Act that gives the administrator of the FAA the power to suspend, modify, or revoke certificates issued by the FAA.

The rules that begin with section 821.30 of Title 49 of the code are extremely complicated and complex. While it is possible for a nonlawyer to attempt her own appeal, a reading of the rules will quickly inform most lay individuals that they should not attempt an appeal to the NTSB without the representation of competent counsel, preferably one familiar with these proceedings.

To begin, the document known as a *notice of appeal* must be filed with the NTSB within 20 days from the time that the FAA serves the order of suspension or revocation upon the respondent. That notice of appeal must contain a concise but complete statement of the facts relied upon and the relief sought. The facts relied upon are generally the facts as the respondent sees them, and the relief sought is most often a dismissal of the order of the FAA. Furthermore, the notice of appeal is required to recite the FAA's action from which the appeal is sought, and it must identify the certificate affected by the order from which the appeal is taken.

One of the most important effects of filing an appeal with the NTSB is that the timely filing postpones the effective date of the order until final disposition of that appeal by either the administrative law judge who is assigned to hear the case or the full National Transportation Safety Board.

However, in emergency proceedings, the effectiveness of the order is not delayed, and *emergency orders* take effect from the day they are issued and served upon the certificate holder whose certificate is the subject of emergency proceedings.

After the notice of appeal is sent to the board and to the FAA, as it is required to be, the FAA has a period of 5 days in which to file a document known as a *complaint*. This complaint is nothing more than the order itself from which the appeal is being taken. These rules of administrative procedure contained in Part 821 have some similarity to the rules of federal procedure that are used in civil lawsuits in the federal courts. Some of the nomenclature is identical, and the word *complaint*, as used in Part 821, comes from the court rules that govern normal lawsuits in the civil courts.

After the FAA files its complaint, which again is the order of suspension or revocation from which the appeal has been taken, the respondent, who is the certificate holder, has 20 days in which to file an answer to that complaint. The answer must deny the truth of any allegation(s) in the complaint unless the respondent admits those allegations. Most often there are allegations in the order (complaint) that can be admitted. Almost always the order will begin its factual allegations by alleging that the certificate holder is, in fact, a holder of the particular certificate in question. Almost never is that an issue in controversy, and it is one that can be admitted. It is a rule of law that a person must deny allegations only in good faith, and a blanket denial of all allegations can sometimes subject the respondent, as well as his attorney, to sorrowful sanctions. In any event, failure to deny the truth of any allegation in the complaint will, in most instances, be deemed an admission of the truth of that allegation.

Also, the respondent must include, in his answer, any defense that he has, which is a type of defense commonly thought of as an *excuse*. These excuses are referred to in the law as affirmative defenses, and they basically say that the action of which the respondent is accused did, in fact, happen but there is a good, sufficient legal reason why that action did not constitute a violation of the FAR involved.

An example is a situation in which the certificate suspension is sought for deviation from an ATC clearance, or failure to comply with an ATC instruction while operating in positive controlled airspace, and the certificate holder is defending on the basis of an emergency situation that required the deviation or that presented the pilot with a condition wherein he could simply not avoid deviating from the clearance or instruction. If any such affirmative defenses do exist, they must be raised in the answer or are deemed to be waived.

It is very important to realize that in appeals before the NTSB, the administrator of the FAA, who is the official person seeking to act upon the certificate of the respondent, has the *burden of proof*. In order to win, the FAA must carry this burden of proof, and prove each and every element of the violation.

The respondent is in a position similar to that of a defendant in a civil lawsuit. To win his case, the respondent need do nothing, if the FAA has not carried its burden of proof. It is only after the FAA attorneys have presented their case, and appear to present a case that is victorious if no defense is mounted, that the respondent must then go forward and present whatever defense he has.

The appeal to the NTSB, in its initial hearing, is not a true appeal as lawyers tend to think of them. Rather, these proceedings are a trial where the administrative law judge is seated to be not only the judge who rules on issues of law as judges normally do but also the finder of fact, which in a typical civil court trial is the province of a jury. Therefore, there is no jury in these proceedings; the administrative law judge serves the role of both judge and finder of fact.

The law judge to whom the case is assigned sets the time and place, as well as the date, of the hearing, and is required to give the parties at least 30 days' advance notice of that hearing. Typically, the hearings are held in a federal office facility—either a federal courthouse or a conference room of a federal office building—at a location that is reasonably convenient for the majority of the witnesses, and with due regard given to the convenience of the parties.

In a typical general aviation case where the incident that gave rise to the certificate action was near the operator's home base, the hearing will be held in the general environment of the respondent's location. However, there are situations wherein the incident might have occurred at some distance from the certificate holder's home, and if a significant number of witnesses are involved, the law judge has the discretion to give deference to the convenience of those witnesses over the convenience of the respondent.

During these hearings, the formal *rules of evidence* as they are applied in court proceedings are not totally applicable. Hearsay evidence is usually admitted in administrative law judge hearings, while it is mostly excluded in courts of law. Practitioners not acquainted with the nuances of NTSB practice need to make a careful review of the standards of evidence admitted and considered by the administrative law judge during the preparation of their cases. Anyone who believes that the strict legal rules of evidence will prevail is operating under a misconception, one that could deal a fatal blow to the respondent's case.

A record is made of all proceedings before an administrative law judge during these appeal hearings. Most often the record is by audio tape recording of the hearing rather than by a stenographic system of reporting, which is generally used in the courts. If any respondent wishes to provide his own stenographic reporter, that is allowed, but the expenses and fees charged by the stenographer are borne solely by the certificate holder who wishes that person's presence during the hearing and who has retained the reporter to be there.

The general order of proceedings used in courts is also followed in NTSB appeal hearings. First, the FAA attorney is given the opportunity to make an *opening statement*, if she chooses, to briefly state what her case is and what she expects the evidence about to be introduced to show in furtherance of the allegations made in the order. The respondent may then make an opening statement, if he chooses. Neither party is required to make such a statement.

At the conclusion of the opening statements, if made, the FAA goes first and presents its *evidence*. Most often the evidence is a combination of oral testimony of witnesses—who might be law enforcement officers, airport personnel, air traffic controllers, or FSDO inspectors who performed the initial investigation—and of other persons who have knowledge of the incident that gave rise to the order of suspension or revocation. Various and sundry documents, including diagrams and drawings, may also be introduced.

When the assistant regional counsel is finished with the presentation of all the FAA's evidence, it is the respondent's turn to deal with what has been introduced. If the certificate holder feels that the FAA has not presented a case that proves a violation, without the respondent's doing any more, a *motion* is appropriate, asking the administrative law judge to dismiss the order and halt the proceedings at that point.

While this motion is usually made to preserve the record for the purposes of further appeal, it is rarely successful. The standard that is applied is simply one that requires the FAA to have proved its case and carried its burden of proof before the respondent says anything or introduces any evidence. It is not a frequent occurrence that the FAA's case is so marginal that it does not present what is known in the law as a *prima facie case*. If the FAA has presented a prima facie case—which is one that, without defense, carries its burden of proof—the motion will be denied and the respondent will be required to go forward and present what defense he has.

At that point, the certificate holder puts on his evidence, which might or might not include his own testimony. He has the opportunity not only to testify himself, but to subpoena other persons and require them to testify, present documentary evidence, and tie together a defense in much the same manner as a defendant would in a court proceeding.

At the conclusion of the respondent's case, the administrative law judge will allow opportunity for *closing argument*. The FAA attorney will go first, because he has the burden of proof, and will argue why the evidence that was submitted in the hearing proves the FAA's case, and why the FAA is entitled to the sanction against the certificate that it is seeking. At the conclusion of the FAA's closing argument, the respondent has the same opportunity, and will likewise argue why the evidence that has been presented, in its totality, does not show that the FAA is entitled to its requested sanction, but rather, entitles the respondent to either a reduced sanction or a complete dismissal of the order.

The FAA attorney is then entitled to a *rebuttal argument* if he so chooses, but this is usually brief, and will respond only to the direct comments made by the respondent or his attorney during the closing argument.

At the conclusion of the arguments, the law judge has a choice as to whether to render an *initial decision* orally or to render it later in writing. This initial decision is required to include a statement of the findings of fact that the law judge has made as a result of the evidence that came out in the hearing, and his conclusions of law that apply the law to those facts and support his eventual decision. In the initial decision, the law judge may do one of three things:

1. Find in favor of the FAA in all respects and uphold the sanction that appears in the certificate action order from which the appeal was taken
2. Find in favor of the FAA but also find mitigating circumstances that justify and, therefore, require a reduction in the requested sanction
3. Find in favor of the respondent, dismiss the order, and impose no sanction upon the certificate holder

One of the truly favorable parts of the appeal process, as far as the certificate holder is concerned, is that the law judge does not have the opportunity to increase the sanction sought by the FAA. Once the administrator's order is issued—and that is, remember, the order of suspension or revocation—that document fixes the maximum sanction to which the administrator is going to be entitled in a later appeal proceeding.

Therefore, the certificate holder is in a no-lose situation when she appeals to NTSB. At the conclusion of the hearing, she could do no worse than the FAA requested in the first place; in many circumstances, if not most, she will likely succeed in obtaining some reduction of the penalty sought in the order of suspension or revocation.

It is unfortunate, but nonetheless true, that because of the appeal option, the FAA often seeks a sanction in an order of suspension or revocation which is higher, or more severe, than it thinks will eventually be given by a law judge if an appeal is taken to the NTSB. For this reason, *most practitioners in this area do appeal these orders*. Even if they do not intend to go all the way through the hearing process, they will use the pendency of the appeal as leverage in negotiations with the FAA and seek a reduction in the requested sanction and possibly compromise and settle the case along the lines of a reduced penalty prior to the actual hearing of the appeal.

This entire process, from the notice of proposed certificate action through the appeal hearing before the administrative law judge, can take a significant period of time. Today's enforcement policy has clogged the system to a rather dramatic extent, and it is not at all uncommon to see a year or more go by between the issuance of a notice of proposed certificate action and the actual hearing before the administrative law judge after an appeal has been filed. After the administrative law judge's initial decision, either party, the FAA or the certificate holder, may appeal that decision to the full National Transportation Safety Board.

Appeals to the full board are a great deal less common than are the initial appeals that are handled by administrative law judge hearings. Appeals to the full board require trips to Washington and extended legal briefs, and are therefore a great deal more expensive than the proceedings just described for the initial hearings before the law judges.

In the event that either party wishes to appeal the law judge's order, it must file a notice of appeal with the full board within 10 days after either the oral initial decision or written decision has been rendered. Within 50 days after the oral initial decision, or 30 days after the written decision, the party bringing the appeal to the full board must file a brief with the NTSB. The appeal will be dismissed if the brief is not filed timely.

The *appeal brief* must set forth, in detail, the party's objections to the initial decision, and must describe, also in detail, whether those objections are related to supposed errors in the law judge's findings of fact and conclusions, or whether the errors are in the judge's order, as such, or both. The appeal brief also is required to state the reasons for whatever objections are made, and the relief requested, which obviously is an overturning of the law judge's decision. If the respondent is appealing, the relief requested is almost always a dismissal of the order; and if it is the FAA who is bringing the appeal, it is usually seeking the relief of having the order of suspension or revocation restored and made effective.

After the initial brief is filed, the opposing party has an opportunity to file a *reply brief*, and that reply brief must be filed within 30 days after the opposing party's appeal brief has been served upon the replying party. No other briefs are generally filed, and no oral argument is usually held before the board. However, if the board feels a need for oral argument, it can permit the same if one of the parties asks for the opportunity to come to Washington and make an oral argument; or, if the board feels an oral argument is necessary, the board has the legal ability to order oral argument without either party's having asked for it.

These appeals to the full board are much more limited than the appeal hearings that first take place before the administrative law judges. The board is considering only the following issues:

1. Whether the findings of fact were supported by legally recognizable and reliable evidence
2. Whether the conclusions made by the administrative law judge were in accordance with the precedent and policy previously announced
3. Whether the questions that are being raised are truly substantial, or whether they are trivial
4. Whether the law judge made any prejudicial errors

If the board rules that the law judge did make a mistake, it has two options. It can make any necessary findings and issue its own order in lieu of the law judge's order, or it can do what is known as *remanding* the case, which simply means sending it back to the law judge with instructions to hold a new hearing along lines dictated by the board in its appeal decision.

After the full board rules, it is possible to request a rehearing, reargument, reconsideration, or modification of the board's order, but all these subsequent attempts are generally unsuccessful.

FURTHER APPEALS

After the National Transportation Safety Board appeal system has run its course, through both the administrative law judge hearing and a subsequent appeal to the full board, and only after those two processes have been completed, an appeal may be taken by either party to the United States Court of Appeals in Washington, D.C.

Only the most serious cases will get appealed this far. At this point, the expense truly increases dramatically, because counsel admitted to practice before that court must be employed. Most often, counsel will be an attorney who physically maintains a practice in the Washington, D.C. area, although several attorneys across the country are also admitted to practice before the District of Columbia circuit.

Appeals to the United States Circuit Court of Appeals are an extremely rare occurrence for general aviation cases. It is usually an airline operator or large air taxi certificate holder who has enough at stake to take a certificate action this far.

After the Circuit Court of Appeals renders its decision, it is theoretically possible to appeal the case further, to the United States Supreme Court. Even a cursory look at the reported cases in administrative law in this area reveals a definite paucity of Supreme Court cases. This is simply because the Supreme Court is not required to hear these appeals, and most of the cases that come before the United States Supreme Court do so only after the court grants permission to hear the cases in the first place. The Supreme Court rarely feels that a case of this type is sufficiently important to occupy the court's time, when considering the thousands of other controversies, each seeking its moment of attention and decision from the highest court in the land.

Therefore, the appeal process beyond the National Transportation Safety Board is extremely expensive, rarely utilized, and seldom successful for the respondent.

CIVIL PENALTIES

In addition to seeking to suspend or revoke a certificate holder's certificate, the FAA has the power to seek a *civil penalty* in the event of a violation of the regulations. A civil penalty is very much akin to what we all think of as a fine in the normal judicial system. A civil penalty case seeks the payment of money rather than the temporary suspension or revocation of the certificate holder's ability to exercise the privileges of that certificate.

In times past, the FAA took the position that a person who uses his certificate in his occupation, such as a mechanic or a commercial pilot or operator, would rather pay a civil penalty for a violation of the regulations than lose the ability to earn his living for a period of time. Again, this is a philosophy of enforcement that is no longer current. Under the former approach, the FAA would assume that those who used certificates in a noncommercial manner (generally private pilots) would rather suffer a suspension of their certificates than part with hard-earned dollars. Then, civil penalties were sought mainly against occupational users of certificates, while suspension was the preferred remedy for the noncommercial user.

Today, civil penalties are sometimes used in a similar manner. When violations occur by airlines, air taxi operators, approved repair stations, and pilots who use their certificates commercially, and the violations are not particularly severe or are not of regulations for which the FAA is making a concerted attempt to obtain rigorous compliance, civil penalties are sought.

Another reason for the FAA's seeking a civil penalty instead of a certificate action relates to the time periods in which each of the two different types of cases must be instituted. A certificate action must be begun, by the issuance of the notice of proposed certificate action, within 6 months of the event which precipitated the enforcement activity. This 6-month time period can be extended for certain reasons, but the certificate holder will hold the FAA's feet to the fire if it takes longer than 6 months to issue the notice.

Civil penalties can be brought up to 2 years after the causative event. Therefore, if the investigation takes longer than normal, or if for any number of reasons or mistakes, the FAA drags its feet and goes beyond the 6-month time period for starting a certificate action, it will opt for a civil penalty in most instances.

In 1989, the procedure for civil penalty cases underwent a complete overhaul from prior practice. For a few years an experimental program was in place, and parts of it are now permanent, while other parts were changed in response to complaints from the aviation industry and press. The new program involves a departure from prior practice in that the courts are not involved in the prosecution or collection of civil penalties, unless the amount sought by the FAA exceeds $50,000. Now, the FAA sends out a notice of proposed civil penalty in much the same manner as when it seeks a certificate action—and very similar options are available in response to that notice—and then issues the civil penalty through an *order of civil penalty*, much like an order of suspension or revocation of a certificate. If the respondent, who is the entity against whom the penalty is sought, is not an individual pilot, mechanic, or holder of other certain certificates, the hearing is held before a Department of Transportation administrative law judge, who is supposed to be considered an impartial decision maker.

Appeal from that decision is to the administrator of the FAA, who is supposed to have been completely removed from the process so far. Much has been written about this new program, and most of it has been extremely critical.

The main reason that the new program was put into effect was that the former program seldom resulted in either prompt or effective disposition of civil penalty cases. Under the old permanent system, before the experimental program was put into effect, civil penalty actions were prosecuted in the United States District Court, in much the same manner as a civil lawsuit proceeds. The FAA would turn over its entire file to the United States attorney within a particular federal court district. When the attorney got around to it, she would file a complaint against the alleged violator and the case would then proceed within the United States District Court system as any other civil lawsuit does. Because most federal courts are extremely busy and docket times between the institution of a case and trial often stretch out into years, this procedure delayed and hampered the civil penalty system.

If the proposed civil penalty does exceed $50,000, or if the government asks the court to enjoin, which in lawyer talk means prohibit, any future activity of the respondent, the case still goes forward in the federal court, the same as it always did.

Also, United States attorneys frequently consider themselves to have far more important things to do than proceed with civil penalty cases on behalf of the FAA. When one realizes that the United States attorneys' offices are actually the federal prosecutors, and deal with such things as counterfeiting cases, drug cases, securities violations, terrorism cases, and similar, serious federal crimes, it is no wonder that they often considered FAA civil penalty cases to be a thorn in their side and a bother, and were not particularly diligent in prosecuting them.

For all these reasons, Congress chose to try a different approach, and authorized the program that now exists. Most of the legal scholars who have studied the new program do not like it, and consider it to be quite possibly unconstitutional.

To appease some of this criticism, the new civil penalty system keeps jurisdiction with the NTSB, just like certificate action cases, if the accused is an individual, as opposed to a corporation or other business entity, and that person is a pilot, mechanic, or the holder of a certain number of other certificates. In that manner, there is no Department of Transportation (DOT) person involved as the judge, since the hearing is held before an NTSB law judge. Likewise, the appeal is to the NTSB, not the FAA administrator.

Regardless of the procedure or amount of the proposed civil penalty, keep in mind that the aviation safety reporting system report can still offer immunity from sanction (your money) if the requirements of the program are met, and you get a report in to NASA within the prescribed 10 days of the incident in question.

I have represented many different entities and people in civil penalty cases of all three types: cases over $50,000 in federal court, cases that went the NTSB route, and cases under the new system in which the FAA administrator is the appellate decision maker. The latter two types of cases, where the penalty is below $50,000, are not handled all that much differently.

But when faced with a huge penalty, the respondent is generally better served by being in a formal courtroom environment. The legal rules of evidence are closely followed and enforced in federal courts, while they are very relaxed in an NTSB or DOT proceeding. Also, the pace of progress in the case will generally be slow in federal court, unless the FAA has sought an injunction to keep the alleged violator on the ground, or to prevent some other action, during the pendency of the case.

Injunctions are sought, as part of a large civil penalty, when the FAA believes that the respondent is likely to continue disregarding the regulations, and persist in some wayward activity. I've seen injunctions used in cases where the defendant was a student pilot, and the FAA had evidence of hundreds of flights where he had carried passengers; cases where an air taxi operation was so shoddy that the FAA wanted to make sure that it ceased operations right away; and similar instances of complete and willful disobedience to the regs.

ADMINISTRATIVE ACTION

Administrative action is a procedure that is least detrimental to the alleged violator. Administrative enforcement action exists to provide FAA field inspectors, within the various FSDOs, a means of disposing of minor types of violations that the FAA feels do not require the use of any legal enforcement sanctions.

According to the FAA's own compliance and enforcement policy, administrative action may be taken only in cases where there is conclusive evidence of a violation, but it does not actually charge the person involved with having violated a FAR. Its purpose is to serve as a warning and, at the same time, to accomplish whatever correction is necessary in the violator's operation to bring her into compliance, and to reasonably ensure that further such transgressions will not occur in the future.

Administrative action basically takes two forms. First, the FAA can issue a *warning notice,* a sample of which is set forth in Figure 6-7. The purpose of the warning is to put the alleged violator on notice of the facts and circumstances of the incident, and to advise him that the operations are contrary to the regulations. Further, the warning notice indicates that the matter either has been corrected or does not warrant further enforcement action. Lastly, it requests future compliance with the regulations.

It is important to know that if a letter of investigation has not been issued concerning the incident in question, the FAA will further invite explanation or mitiga-

November 20, 19— Case No. 88WM010000

Mr. Fred Smith
1075 Victory Boulevard
Los Angeles, California 90009

Dear Mr. Smith:

On October 20, 19—, you were the pilot in command of a Beech Baron N13697 that landed at the City Airport. At the time of your flight, you did not have in your personal possession a pilot certificate, although you do hold a valid commercial pilot certificate. This is contrary to the Federal Aviation Regulations.

After a discussion with you concerning this flight and your inadvertent failure to have your pilot certificate with you, we have concluded that the matter does not warrant legal enforcement action. In lieu of such action, we are issuing this letter which will be made a matter of record for a period of two years, after which the record of this matter will be expunged.

If you wish to add any information in explanation or mitigation, please write me at the above address. We will expect your future compliance with the regulations.

Sincerely,

JOHN J. FRANK.
Chief, Van Nuys GADO

Attached: Privacy Act Notice

Fig. 6-7. Sample warning notice—flight operations.

tion from the recipient as a part of its warning notice. A warning notice is not final, and can be withdrawn if the FAA decides that legal enforcement action is appropriate. Therefore, beware. Do not volunteer information that is at all incriminating, and treat the warning notice with care. Once a warning notice is issued, it is rare that it will be withdrawn and the case converted to one of legal enforcement through either civil penalty or certificate action, but the possibility exists, and everyone must be aware of that.

The second form of administrative action consists of a document known as a *letter of correction*, a sample of which is shown in Figure 6-8. The letter of correction serves basically the same purpose as a warning notice, but the circumstances under which it is used vary slightly. The warning notice is used to put the alleged violator on notice that a violation has allegedly occurred. The letter of correction is used when there is agreement with the pilot, company, or other organization that a particular corrective action acceptable to the FAA has been taken, or will be taken within an agreed-on period of time.

When a letter of correction is to be used, therefore, there is usually some discussion between the alleged violator and an FAA representative that results in a plan of correction of a minor violation. The letter of correction confirms this discussion and sets forth what will be done to correct the matter at hand.

If a plan of correction is not followed by the violator, the FAA will institute more severe enforcement action, which will usually take the form of either a civil penalty or a certificate action. One cannot treat a letter of correction lightly, because follow-up will occur, and the recipient must keep the bargain and correct the situation that led to the letter.

April 30, 19--

The Aerospace Company
Attention: Mr. J. A. Jones, President
1200 International Way
Newark, New Jersey 22180

Dear Sir:

Your repair station's organization, systems, facilities, and procedures were examined for compliance with applicable Federal Aviation Regulations (FAR) during the period April 1-10, 19--. At the end of that examination, you were advised of our findings as follows:

The summary of employment of each person whose name appears on the roster of supervisory and inspection personnel was not available for three of the employees, as required by Section 145.43(b) of the FAR.

This is to confirm our discussion with you on April 8, 19--, at which time immediate corrective action was begun. You submitted the required summary of employment for FAA inspection on April 10, 19--.

As a result of our discussion of this incident, you have revised your procedures for maintaining the required summaries of employment.

In closing this case, we have given consideration to all available facts and concluded that the matter does not warrant legal enforcement action. In lieu of such action, we are issuing this letter which will be made a matter of record.

Sincerely,

JOHN L. DOE
District Office Chief

Fig. 6-8. Sample letter of correction—maintenance.

If the administrative action is against an individual as opposed to some business organization, the record of the administrative action will be purged after a period of 2 years following its issuance. As to businesses and certificate holders who are not certificated pilots, the administrative action remains a matter of record indefinitely.

We have seen a resurgence in the use of administrative action, because the FAA has realized that certain parts of its very strict compliance and enforcement program were too rigid, and lacked sufficient flexibility to accomplish what everyone realizes is the end goal—improved aviation safety.

Many first offenders find themselves within the grip of the legal enforcement program resulting from situations in which they were actually trying to comply with the regulations, but through ignorance, noncurrency in their training, or an honest mistake, they fell short of the mandates of one or more of the FARs. More administrative action is being used for first-time offenders of the FAR sections that deal with airspace and airspace restrictions, particularly operating in Class B and C airspace areas without strict adherence to the regulatory requirements.

The FAA adopted a program which is known as remedial training. If an inspector who is assigned to the case feels that the pilot committed a violation through ignorance, and if the person involved demonstrates an attitude of being desirous of compliance with the regs, the inspector may, but is not required to, divert the matter from formal enforcement action into the remedial training program. The pilot will be counseled by the person within the Flight Standards District Office who is responsible for that office's aviation safety program. The counseling will be aimed first at determining the pilot's shortfalls and deciding what type of additional training might help prevent such problems in the future.

Then, a specific agenda of additional training, usually including both ground and flight sessions, will be developed. The pilot and the FSDO safety manager will agree on a particular flight instructor or other training facility to provide the required training. When it is completed, a report will be sent back to the FSDO that the training was satisfactorily accomplished, and the matter ends there. If the pilot does not complete the training, or if for any number of other reasons the instructor doesn't report favorably to the FSDO, the FAA still has the option of proceeding to formal enforcement action.

If you are ever the subject of potential enforcement action, bring up the remedial training program to the FSDO inspector, either personally or through your attorney. If you meet the requirements for this diversionary program, and most particularly if you demonstrate a compliance-minded attitude, you may well save yourself from a suspension or a civil penalty. Most important, you'll learn something in the process.

This remedial training program is definitely a good sign, and in my humble opinion, it will result in greater safety. After a violator has gone through the legal enforcement system and suffered either a civil penalty or certificate action, there is no guarantee that, as a result of that process, the violator understands what he did wrong, or has learned what he needs to know to prevent further violations.

Because drastic changes have occurred in the airspace system and structure within the last decade or two, there are many certificated pilots who were trained in a simpler time and who do not venture with any regularity into complicated airspace. Even though the biennial flight review is supposed to address this problem to some degree, it is certainly not a cure-all. If minor violations of airspace are treated as situations that need educational redress rather than legal enforcement action, those pilots who sincerely want to comply with the regulations will be equipped with the tools to do so. Therefore, administrative enforcement action might well continue to be used, in appropriate cases, as an alternative to formal enforcement action.

7

Principles of Negligence Liability

WITH THE ADVENT OF CIVILIZED SOCIETIES, THE PEOPLE WITHIN THEM began to realize that in order to live together, particularly in an organized fashion and in close proximity to one another, each person must be held accountable for what she does and does not do when an action, or inaction, affects another person. Sanctions for failure to conduct oneself appropriately run the entire gamut of legal theories and remedies.

The state has laid down, in its body of criminal law, a large collection of statements of what individuals and businesses may not do. When a person runs afoul of those mandates, the state will come forward and prosecute the offender, seeking either to incarcerate him for a period of time or to obtain monetary retribution in the form of a fine, or perhaps try to exact both means of punishment. The government prosecutes criminal violations in the name of society at large. In some states, criminal cases are formally presented to be a situation where the people of that state are proceeding against the defendant who has been accused of violating the criminal law.

In the area of *negligence liability*, generally one individual is seeking redress for a legally recognizable wrong done to him by one or more other persons. Remember, the word *person* refers not only to individuals but also to the other entities and collections of individuals that are recognized as persons in the law. Such additional persons, beyond human individuals, include corporations, unincorporated associations, and governmental bodies and entities.

The very heart of negligence law is the concept that we must act, as civilized people, with a view toward how our conduct affects others. Negligence has been recognized as a legal wrong for centuries. Although many other areas of the law have been as well, this principle has seen profound evolution and development over the last few decades. The concept of negligence involves primarily four elements:

1. One's duty to others
2. Whether that duty was breached

3. Whether that breach of a legally recognizable duty was the legal cause of the harm of which the injured party complains
4. Whether any legally recognizable damages resulted from the breach of the duty owed to someone else

Each element will be discussed in detail in order to first understand the general principles of negligence liability prior to application in Chapter 8 to particular aviation situations.

DUTY TO OTHERS

Before any other element of the wrong of negligence can be discussed, we need to examine the foundational block upon which the tort of negligence depends. That foundation is the finding that, under the situation at hand, the person being accused of negligence had a legal *duty* to do something, or not do something, which is at issue in the case. The body of tort law recognizes many duties that have been developed over the centuries by the courts through the common law system. In addition to those duties that have been announced by judges, the various legislative bodies throughout this country have, through enacting ordinances and statutes, dictated certain duties that each of us owes to our fellow citizens.

For instance, in the area of motor vehicle law, speed limits exist for several purposes. When we violate the speed limit law, we can come under the scrutiny of the criminal system if a law enforcement officer observes the speeding and chooses to issue a citation for what is a criminal violation of a traffic law. When that occurs, we are thrust into the realm of a criminal violation; but a simple violation of the speed laws and the resulting speeding ticket do not result in any private right of another individual to sue us in a courtroom for monetary damages.

The very essence of a *negligence action* is that a person is being sued, in a civil court, by an injured party who is seeking monetary damages from the defendant. That injured party then must maintain and prove that, first of all, the defendant did not act toward that plaintiff in a manner in which the law required her to act.

Let us take the example just discussed, a simple act of speeding, and change the facts slightly. Instead of being pulled over by the local police officer with red, blue, or white lights flashing, our driver is speeding while an automobile down the road is pulling out of a parking space into the normal flow of traffic. When the driver pulling out of his parking space looks to his rear, he sees the car coming, but reasons that there is a sufficient distance between the oncoming vehicle and his own, so that he will have no problem pulling out of his parking space and entering the flow of traffic. The driver of the car entering traffic is certainly entitled under the law to assume that the oncoming vehicle is obeying the speed limit. Unfortunately, what the driver pulling away from the curb is unable to discern is the fact that because the oncoming vehicle is exceeding the speed limit, the closing time until impact will be greatly reduced from what is justifiably assumed. When the speeding car then impacts the one coming away from the side of the road, the speeder has breached her duty. She has caused an accident because she did not do that which she is supposed to do—drive her car at or below the speed limit.

There are many permutations to such an example. If the driver pulling away from the curb had opportunity to view the oncoming vehicle for a sufficient period of time to determine that it was speeding, then perhaps he would not have been justified in assuming that the other driver was operating within the speed limit. But the concept is the same. Each has a duty to obey the traffic laws and conduct himself within them, exercising reasonable care to the other individuals occupying the public highway. When a driver fails to do that, he has not conducted himself in accordance with the duty that the law requires.

In most circumstances, the law requires each person to exercise *reasonable care* for the safety of others. This duty of reasonable care is one that operates in a legal fiction, but does have specific bounds and definitions. In most states, and in most situations, the courts use a standard which is hypothetical, but to which lawyers, judges, and legal writers have become accustomed over the period of time since the common law first began to develop. That fiction is known as the *reasonable man standard.*

The law looks at a particular situation and asks the jury for a determination: "What would the reasonable man do in similar circumstances?" Therefore, we are not limiting our inquiry solely to what the particular defendant did in the circumstances, but we are, instead, comparing the defendant's conduct to what a reasonable person would do in similar conditions.

Another good example can be drawn from the area of highways and automobile driving, as most of us are certainly familiar with that endeavor. Let us say the speed limit is 65 along the interstate in a rural area of Ohio, as it now is since the federal government eliminated the absolute requirement for a 55-mile-per-hour limit. But perhaps the particular evening in question is a dark, snowy, windy night in January, at which time there are already 3 inches of snow on the ground, and heavy snowfall is still occurring.

Even though the speed limit is 65, I doubt that many people would argue that the reasonable person would continue to operate a passenger car down a rural freeway under such conditions at 65 mph. Therefore, if an accident occurs on that freeway under those weather conditions, no doubt a question will be presented to the jury as to what the reasonable person would be doing, assuming he would even be on the road in the first place.

Without question, the reasonable driver would be operating his car at a speed substantially below 65 mph; but it is up to each particular jury in a lawsuit to determine what it feels the reasonable person would do, and then compare the defendant's conduct with that standard. This is what is meant when the concept of reasonable care as exercised by the reasonably prudent person is discussed.

There are situations where the law requires a higher duty than merely reasonable care. For instance, people receiving the property of others to hold for a period of time for charge or other compensation are held by the law to a duty higher than ordinary care. There are other scenarios in which the law recognizes that one owes the highest possible standard of care, such as the duty owned by an air carrier to its paying passengers. Each of these various duties is discussed in detail in Chapter 8 as the duties change, elevate, and decline, depending upon the circumstances and facts in certain aviation operations.

The law does recognize the possibility of an unavoidable accident, and therefore we cannot lose sight of the fact that, in order for a defendant to be *negligent*, the first and foremost inquiry is whether the person being sued had a duty to act or a duty to refrain from some activity. This duty must be a legally recognizable duty, as opposed to one that is merely compelled by a sense of morals. For example, the law in most states does not require a child to render care to his ill parent, unless under very certain and limited circumstances the parent is indigent and about to become a burden upon the taxpayers, and the child is an adult. But if a person's parent is suffering from a bout of the winter flu, and let us assume lives across a small or medium-size town from his adult child, the child has no legal duty to go to the parent, provide care, fix meals, tend the house, or do similar chores or gratuities. Most of us would recognize that as a moral duty, but a clear distinction must be made between a duty legally required and one that is simply indicative of how people ought to act toward one another when they are not necessarily legally obligated to do so.

In a negligence case, it is crucial that the plaintiff establish a legal duty that the defendant had toward that plaintiff. That is not limited to a duty to do a particular thing, but can be a duty to refrain from doing something. A hunter has a duty to exercise caution when he fires his weapon, and to not fire it in an indiscriminate manner so as to injure another person, livestock, or other property of the owner or occupant of the land upon which he is conducting his hunt. If the hunter flushes a pheasant and fires his shotgun at it, and in the process shoots a cow, he had the duty to refrain from shooting the livestock. Assuming all the other elements can be met, the hunter would then be guilty of negligence toward the owner of the cow. In a similar manner, we have the duty to refrain from slandering our neighbors, libeling them, or visiting upon them a host of other miseries.

The reader can easily see that there can be a myriad of situations wherein it would be difficult to determine whether the duty is one to act in a particular manner or, conversely, not to act in a certain way. The law does not allow such confusion to stand in the way of a jury's finding of a duty. Whether the duty is one of commission or omission, if the jury finds that, under the circumstances of a particular situation, a duty existed on the part of the defendant to act or not act in a specific way toward the plaintiff, and that duty is one that the law recognizes, the plaintiff has then carried her burden of proof and established her case as to this one element of negligence.

BREACH OF DUTY

Once the plaintiff has proved that the defendant owed a particular duty to her, she must next prove that the defendant did not comport his conduct to that duty, and thereby breached it. The word *breach* is used often in various legal theories that are not limited to tort liability, such as warranties and controversies involving contracts. Basically, a breach of a duty is simply not performing as the duty requires. If the court finds the existence of a particular duty in an automobile case involving the running of a stop sign, the jury must next find that the defendant did not stop at the stop sign in the first place, or did not remain stopped until such time as a conflict in traffic was no longer a factor, under circumstances where this was required.

The same analysis will apply to aviation cases as well. If the jury finds that a mechanic with inspection authorization, during the conduct of an annual inspection on an airplane, has a duty to inspect a portion of the airplane and observe a condition that requires repair, the jury must next expand its inquiry into whether the mechanic did in fact make that inspection and observation. A breach of the duty to inspect could occur if a mechanic fails to inspect the area at all, or even where a mechanic performs the inspection but fails to properly observe and uncover a dangerous condition.

Once those questions are answered, the plaintiff would seek to show that the mechanic had a further duty to either repair the defect in question or, if the aircraft owner did not authorize the repair, bring it to the owner's attention and, if it involves a problem that renders the aircraft unairworthy, declare the airplane to be unfit for flight and not return it to service.

So the mechanic in this scenario could breach his duties in several ways:

1. He could fail to inspect
2. He could inspect but fail to uncover a dangerous condition
3. He could inspect and uncover the dangerous condition but fail to repair it properly
4. He could do all the foregoing and not bring the unrepaired defect to the owner's attention and return the aircraft to service in an unairworthy condition

In most cases, when a duty is identified, significant argument and controversy will exist over whether that duty was breached. Not all cases are as simple as whether someone stopped at a red light or a stop sign, and when aviation cases come to the attention of the court, they seem to become much more complicated. Perhaps this complication arises because of the nature of the beast. Most judges and almost all jurors have very little, if any, exposure to aviation matters, while everyone involved drives automobiles and can much more quickly and thoroughly understand the issues involved in a run-of-the-mill automobile accident case. Furthermore, the obvious dangers that result from breaches of duties in aviation cases are quickly seized upon by courts.

I once tried a case involving an accident that occurred after an ordinary fly-in that was conducted at a rural county airport. The accident involved an aircraft that had not had an annual inspection in years, and was owned and flown by a man who at the time had absolutely no pilot certificate whatsoever, but was rumored to have possessed a student pilot certificate two or three decades prior to the accident. He was carrying a passenger, and performing what was then commonly referred to as a bomb drop run during the fly-in. The purpose of the maneuver was to fly over the runway at a relatively low altitude and slow airspeed, and drop a sack of flour on a target. Many aircraft participated in the exercise, the winner was the crew who was able to plant their flour sack closest to the target, which was laid out in the grass alongside the single runway.

After the bomb drop run was completed, during the go-around, the aircraft stalled a few hundred feet above the runway, entered either a very steep spiral or a spin, and crashed off the end of the airport. The passenger was killed in this accident, and the pilot suffered serious injuries, but survived and recovered to a nearly full extent.

The passenger's estate filed suit against the pilot, as would be expected, and somewhat unexpectedly, also pressed a claim against the fixed-base operator who ran the airport. Unfortunately, the pilot was totally uninsured and not a man of significant financial means. Prior to the commencement of trial, a settlement was reached between the pilot and the plaintiff passenger's estate, and the only defendant remaining for the trial itself was the fixed-base operator.

The plaintiff advanced the position that the airport operator had a duty to ensure that those aircraft based at his field were airworthy and maintained in accordance with the FARs, and had a further duty to ascertain that pilots who flew regularly from the field were properly licensed to engage in the operations involved.

The defense took the position that an airport operator had no such duty, but rather, in terms of the facts involved in this particular case, was nothing more than one in the position of a parking lot operator who simply rented storage space to this gentleman for the tie-down of his aircraft, much in the same manner that a parking lot rents space on a monthly basis in urban areas for those people who work in the city to park their automobiles.

The court readily agreed that such a parking lot operator had no duty to make sure that those people who parked cars in a garage had their cars properly maintained and registered with the state, and the court also agreed that such a garage operator would not be responsible to establish that drivers of those cars possessed valid drivers' licenses, drove their cars properly, were not intoxicated, and similarly conducted themselves in accordance with the requirements of the law as it affects auto owners and drivers. However, when presented with the analogy that the defense put forth in this case, the court remarked, "But this is an airplane case, and that is different."

Therefore, one needs to be thoroughly aware that aviation cases are dealt with by the courts as situations that have a high potential for danger, and the law will most often scrutinize aviation considerations to a much higher degree than it will activities in which the average person engages and with which average people are much more intimately familiar. Perhaps this would not be the situation in an idealistic world, but we must realize that judges are people, and will bring foibles with them to court.

In the trial of this particular case, a settlement was reached between the insurance company for the fixed-base operator and the deceased passenger's estate during the conduct of the trial. Therefore, no judicial result occurred, and no opportunity was given for any legal precedent to arise out of this rather unusual case.

By looking at these and similar cases, it is quite easy to see that the establishment of a duty and the concomitant (accompanying) breach of it are not simple issues, and are quite often determinative of the outcome of a lawsuit. Whenever one or two particular issues will weigh heavily in the result of a court case, an intense amount of effort will be expended by both sides in arguing their relative positions and trying to convince the court and the jury either that the duty did or did not exist or that a breach did or did not occur.

PROXIMATE CAUSE

Once a duty has been established, and further, the court has found that the duty was breached, the analysis of whether the defendant was negligent toward the plaintiff is not finished. The next question that must be addressed and resolved is whether the breach of the duty established was the *proximate cause* of the harm that the plaintiff claims to have suffered.

The concept of proximate cause has often escaped exact definition. It is most closely related to the idea of being the direct cause. Again, for the sake of simplicity, let us use an example from an automobile accident case.

Assume that a driver is approaching an intersection equipped with a functioning traffic light signal. The driver, who is coming toward the intersection, has the red light and is required to stop until the light turns green. Therefore, the establishment of a duty is clear. Further, our scenario will involve the fact that the driver approaching the red light does not stop, but continues on and sails through it, oblivious either to its existence or to his duty to come to a stop and wait until the signal changes, allowing him to proceed. In other words, our driver runs the red light. As he enters the intersection, he slams broadside into an automobile coming from his right, and caves in the driver's door. That results in severe impact and contusions to the body of the person proceeding rightfully through a green traffic signal. So far, it seems like a reasonably simple case. Let us complicate it further by the fact that when the victim's automobile comes to rest, the driver is found lifeless in her seat.

Still, things seem to be fairly simple. One driver ran a red light, resulting in a severe impact to the driver's side of a car that was rightfully proceeding down the highway, and unfortunately an innocent victim had lost her life. Now, let's reveal one additional fact. The deceased driver is taken to the morgue, where a complete and thorough autopsy is performed on the body. The pathologist performing the autopsy comes to the conclusion, which is extremely well supported by all the evidence uncovered in the autopsy, that moments before the accident, the deceased driver suffered a massive myocardial infarction, commonly known as a heart attack, and was either dead or in the process of dying at the moment of impact. Does this state of affairs change the situation any?

Certainly it would. In order for the driver who ran the red light to be legally responsible for the death of the person who was lawfully operating her automobile, that wrongful conduct of running the red light must be the cause of the harm suffered by the victim. Assuming that the evidence can be established in the manner in which it has been set forth here, the jury would most likely find that our victim had already met, or was in the process of meeting, her demise with no interference from the collision that resulted from the other driver's having run the red light. Hence, in this somewhat convoluted scenario, the wrongful conduct on the part of the one who failed to stop at the traffic signal had no causal effect to the harm suffered by the other, and therefore, the law would hold that the otherwise negligent conduct of the driver who proceeded through the red traffic signal was not the direct or legally recognizable proximate cause of the death of the victim.

In order to find that the wrongdoer is legally negligent toward the plaintiff, the conduct of which the complaint is made and around which the lawsuit revolves must have this direct causal relationship to the harm suffered by the plaintiff. Without that, even though the conduct is otherwise wrongful, it will not render the defendant responsible to the plaintiff in a courtroom, and the defendant should, therefore, win the trial.

DAMAGES

After the plaintiff has proved a duty—the breach of the duty—and the fact that the breach of that duty was the proximate cause of the injuries, either bodily or monetary, of which the plaintiff is complaining, his case is not yet finished. Next, the plaintiff must prove to the jury that he has suffered *damages* that again, must be legally recognizable, before he can consider his case a success and recover from the defendant.

Damages, in most accident cases, are comprised of two elements. The first is called special damages, and the other is known as general damages.

Special Damages

Special damages is the term applied to the actual expense or out-of-pocket losses caused to the plaintiff by the wrongful conduct of the defendant. The first component of special damages is the expense that relates to any physical damage to property as a result of the negligence of the defendant. In the case of an aircraft or automobile accident, this is obviously the cost to repair the airplane or car, if repairable, or the total value of the property at the time that it is completely destroyed in the accident. If the vehicle is damaged to an extent that it can be repaired, in most instances the plaintiff can also recover for the loss of use of it during the time that it reasonably takes to accomplish the required repairs and return it to service. To the contrary, if the aircraft or auto is destroyed in the accident, the defendant is usually liable only to respond to the extent of the actual market value of the property at the time of that destruction, and is not liable for lost use because, once the property is destroyed, there is no use available that can be lost.

Often the law takes what are practically unreasonable approaches and certainly no one, including this writer, maintains that our legal system is perfect. When an aircraft or other property is destroyed in an accident, the law assumes that the owner is immediately capable, both practically and financially, to replace it, and then seek the cost of that fair market value of the damaged property from the wrongdoer. Although this is seldom the case unless very adequate insurance is present, the law does not permit the owner of destroyed property to collect for lost use after the total destruction.

Special damages include medical expenses directly relating to physical injuries received by the plaintiff in whatever accident has occurred. Medical expenses include not only the charges by physicians and other direct health-care providers, but also charges for hospital care, prescription drugs, any particular appliances that might be necessary such as braces, casts, and hospital equipment that had to be purchased or rented during a recuperative period, and similar items that directly relate to the provision of medical care to a person who was injured as a result of the defendant's negligence.

Most of the time someone who is injured will also have a period of disability during which she is unable to earn her livelihood. Lost income or lost wages are available as an item of special damages, if they exist and are provable. It is in this area where many plaintiffs run into problems of the provability of their claims. Self-employed individuals often have difficulty in establishing, to the certainty required by the law, what their actual lost income is from a relatively short period of recuperation from injuries received in accidents.

There are some self-employed individuals—doctors, lawyers, accountants, and others whose main source of income is directly derived from their professional services—once they are unable to render those services for a period of time, who find it not particularly difficult to historically establish the value of a day, week, or month of work, and therefore meet the requirements to collect that lost income. Then, there are those self-employed individuals who are engaged in mercantilistic endeavors, such as store owners and salespeople, whose income is not directly related to the amount of time that they spend at their daily endeavors. Those individuals have a more difficult, but certainly not impossible, time establishing the exact extent and nature of the loss they suffered to their ability to earn income during a period of temporary disability while recuperating from injuries received in an accident.

The last general component of special damages covers any other expenses of an out-of-pocket nature. They include travel expenses to receive medical care or therapy, and such usually minor items of expense as clothing or luggage that is damaged or destroyed in an accident.

General Damages

After the plaintiff proves his special damages, he is then entitled to go on and attempt to establish that he suffered *general damages* as well. The largest component of general damages in most cases is that sum being sought by the plaintiff to compensate him for the pain and suffering that resulted from the negligence of the defendant. It is most difficult to place a precise amount on anyone's pain, suffering, and general inconvenience from having to undergo injury, medical treatment, and a recuperative process that can sometimes stretch out for a significant period of time.

In most cases, the amount that a jury will award for pain and suffering bears some relationship to the amount of the special damages that are proved for medical treatment and lost wages. These items of lost income and medical expenses determine, in most people's minds, how severely the plaintiff was, in fact, injured, and give the jury the ability to come to a reasoned conclusion as to the amount of money that the plaintiff should be awarded to compensate her for her pain and suffering.

It used to be said that pain and suffering awards were often arrived at by using a multiplier of the special damages for medical expenses and lost income. Sometimes juries still intuitively do that, although in recent years the process has become less exact. It is possible to have significant problems from an injury that do not require a great deal of medical expense. Probably the best example of that is a situation where an injured party is rendered either a paraplegic or a quadriplegic as a result of an accident. In most paraplegia cases, there is very little medical treatment that can be done, and quite often the period of hospitilization following the actual accident is relatively short, considering the devastating effects upon the remainder of the life of the victim. In such a case, the plaintiff will certainly seek, and, assuming all the other elements of negligence are shown, the jury will most likely award, an amount for pain, suffering, and permanent impairment of life that will be far in excess of the amount spent for medical care following the injury.

Conversely, there can be severe injuries that result in extremely high medical expenses but from which the patient is fortunate enough to completely recover, with no lasting effect upon his ability to earn income, live a normal life, and continue relatively free of long-term pain or other types of suffering. In this situation, the jury might well render a verdict for pain and suffering that does not bear a multiple relationship to medical expenses to anywhere near the degree that would be seen in a situation where the plaintiff was permanently paralyzed.

Although touched upon in the preceding paragraph, another element of general damages is any permanent impairment that the plaintiff might suffer as a result of the injuries received in the accident. The most obvious permanent impairment is the situation where some degree of lifetime paralysis results, and the devastating effects that stem from it. But certainly less severe types of injury can impair a person's ability to do the things that he was able to do before the accident. Perhaps a broken back, although healed and not resulting in any paralysis, might restrict the ability to engage in athletic endeavors, lift heavy weights, stand for extended periods of time, or do other such things that are not immediately apparent to those of us blessed with excellent health.

Where permanent impairment is a fact that can be proved, the jury will certainly take it into consideration and adjust the general damage award accordingly. This area is sometimes abused by plaintiffs who claim permanent impairment from what appear to be relatively slight injuries. Somehow, the common sense of the jury will usually see through a half-baked claim that one can no longer engage in such activities as golf, lawn mowing, housework, and other everyday doings when the injury has totally healed and the medical expenses are slight.

If the plaintiff is a married individual, most states allow the spouse to recover for the losses he or she suffers as a result of the injuries visited upon the spouse. This item of general damages is referred to in most jurisdictions as *loss of consortium*. Consortium includes many of the things that spouses provide the other, including but not limited to sexual contact, the ability to be a devoted husband or wife, the ability to perform normal household services such as maintenance, lawn mowing, laundry, housecleaning, and other such activities that most individuals perform around their places of abode.

A good many states, however, do not permit consortium claims from the children of injured parents, and do not allow recovery for the diminution in the ability of an injured parent to perform the activities that he or she was able to do with and for the children prior to the accident. Moreover, there is a force in the law that is moving to change this fact in most states when the victim's death results from the accident in question. This theory recognizes the fact that the loss of a parent by a young child is devastating, that the remainder of the child's life will never be the same again, and that some degree of compensation ought to be paid for that tragedy.

If the plaintiff in a particular negligence case is the estate of a victim who has met his demise in the accident, the rules of damages change significantly.

First of all, the estate must prove what economic loss it suffered as a result of the premature death of the decedent. In the situation where the decedent was gainfully employed, economists must be retained to testify concerning the financial repercussions involved. This would include details regarding the income of the decedent, the remaining work life period that he or she would have been expected to enjoy, the increases in income that would have occurred during that remaining work life expectancy, calculation of such things as fringe benefits provided by the employer and governmental fringe benefits, all of which form a total package of economic loss to the family resulting from the untimely death of the accident victim.

After that calculation is complete, almost all the states require that the economist then reduce it to present value. I know of at least two states that allow juries to render gross verdicts, wherein the actual economic loss is simply multiplied by the remaining work life expectancy, and such a gross award is then given to the estate. It must be remembered that almost all states do not follow this aberration, but require a reduction to present value. The argument then ensues between the economists for the plaintiff and the defense over not only the components of the calculation, but also the discount rate used to reduce the gross economic loss over the remaining work life expectancy to a present value to be argued to the jury.

After that economic loss is determined, the estate is permitted to recover for any medical expenses charged during the time that medical treatment was rendered to the decedent prior to his death. Obviously, this number can either be large or small, depending on the amount of time involved between the accident and the commencement of medical treatment, and the death of the accident victim.

In death cases, the general damages are considered much differently than they are in the situation where the victim survives and is presenting her own claim and, perhaps, a claim by her spouse for loss of consortium. When the estate of a deceased victim is suing the person allegedly responsible, under negligence law, for the accident, the amount which that estate can recover for noneconomic loss varies widely among the states.

A few decades ago, it could comfortably be said that most states limited the recovery in death cases to the economic loss suffered by the estate, and the damages portion of a wrongful death trial was mainly an argument over arithmetic concerning the work life expectancy, the expected income during the expectancy, and the calculations resulting in the reduction to present value. But in the last few decades, a majority of the states have expanded the opportunity for the estates of deceased victims to seek compensation for general damages suffered by the remaining next of kin.

In most states today, an estate can present evidence and seek to recover damages for the loss of society of the decedent, the loss of potential inheritance that would have been received by family members had the decedent lived out his life, the loss of companionship that family members would have enjoyed had the untimely death not occurred, the loss of support that the family could have expected to enjoy, the loss of services of the decedent, and the mental anguish incurred by the surviving spouse, minor children, parents, or next of kin of the decedent.

There is no question that the development of the law in the area of wrongful death has greatly increased the exposure that defendants face when sued over an accident in which the plaintiff's decedent has been killed. In several cases, these items of general damage are at least as significant, if not more so, than the amounts of actual economic loss suffered by the family as a result of the death of a breadwinner.

Many writers and commentators have lamented at length over the past several years about the increasing number of liability settlements and verdicts being rendered by the courts. It is not the function of this work to make a social or political statement about the amounts to which plaintiffs ought to be entitled and what the law ought to be, but rather to set forth what it is. In almost all instances, it has been the elected legislators serving in the statehouses of the capitals in this country who have chosen, in the exercise of the political and legislative process, to expand the areas of recovery available to plaintiffs. This is not an area of judge-made law, but one of legislative statute. Whether the result is good or detrimental is, and must be, left to the wisdom of those processes.

If a person disagrees with the result, her redress is to the legislative body that created the current state of the law. All too often the courts, juries, and lawyers are publicly blamed for what is actually the work of legislators working in a political environment.

Preponderance of Evidence

Regardless of the nature and types of damages that the plaintiff has suffered as a result of the negligence of the defendant, these damages must be proved by a preponderance of the evidence to the same extent to which the plaintiff must prove the issues of duty, breach of duty, and proximate cause.

The concept of *preponderance of the evidence* relates to the amount of proof that is required in order for the plaintiff to win his case. A preponderance is defined as a quantity of evidence that is credible, reliable, and believable by the jury, and that, in its totality, is greater in weight than that evidence produced by the opposing side. To further elaborate, the legal concept of preponderance implies that one side will produce more evidence than the other, but more is all that is required.

The concept of preponderance as used in civil courts in these types of cases can be perhaps understood by comparing it with the amount of evidence that is required in criminal court for the prosecution to gain a conviction of a defendant who stands accused of a crime. The criminal law requires proof beyond a *reasonable doubt*. That is an amount of proof that, when taken as a whole, leaves no reasonable doubt in the mind of the jury that the crime has been committed, and that the defendant is the one who committed it. Certainly, the law never requires proof beyond all doubt, and embedded in our legal history is the principle that in the criminal arena, what is required is the quantity of proof that renders the defendant guilty beyond all reasonable doubt.

In civil cases, where only money is at issue, the law imposes a far lesser burden of proof, which is the standard of preponderance just discussed. When the law seeks to incarcerate a criminal defendant and deprive him of his most precious possession, his liberty, the higher standard of proof beyond a reasonable doubt must be met. To the contrary, when the law is being used to deprive a defendant of money, regardless of the amount, the courts require only that the plaintiff produce more evidence than the defendant, and the system does not require that the plaintiff prove her case beyond all reasonable doubt.

Again, this is an area of the law that has been steeped in the traditions and pronouncements of the common law for centuries, and is one that will undoubtedly be with us for the foreseeable future.

DEFENSES TO NEGLIGENCE

When the plaintiff has made a case by introducing evidence that proves, by a preponderance of the evidence, that all the elements of negligence that have been discussed above are present, and that the defendant is liable to him, all is not lost to the defense. There are defenses available to obviate the otherwise present liability that the defendant would face toward the injured party.

Even though the defense has been unable to convince the jury that a particular duty did not exist, that the defendant did not breach the duty, that that breach was not the proximate cause of the harm, and that the plaintiff did not suffer the claimed damages, there are still other defenses available. Chapter 5 touches upon the concept of an *affirmative defense*. An affirmative defense is one in which a legally recognized excuse is offered to relieve the defendant of liability. Affirmative defenses are theories that the law has developed over the centuries that, in essence, say that yes, the injury did occur, but there is a reason that this defendant should not be held liable to the plaintiff. A classic affirmative defense of which we are all aware comes from the criminal law: *self-defense*. In a self-defense case, there is an injured or dead person, and those facts are easily proved by the prosecution. However, even though a death has occurred, if the defendant can prove the elements of self-defense, the fact that he killed the decedent is legally forgiven. That is an affirmative defense: "I did it, but I have a legally recognized excuse." There are many affirmative defenses, but the most common ones in the civil law are assumption of the risk, contributory negligence, comparative negligence, and last clear chance.

Assumption of the Risk

Assumption of the risk is a defense wherein the defendant puts forth the theory that the plaintiff knew of the risk of harm to him that was inherent in the suggested activity and voluntarily assumed that risk. This theory of defense implies that there is a risk of harm that is inherently present in the particular activity that gave rise to the lawsuit, and that the plaintiff knew of that risk and voluntarily exposed himself to it, thereby assuming it. If these elements of assumption of the risk can be proved by the defense, the jury might find that the plaintiff did, in fact, assume the risk of harm and exonerate the defendant from liability.

An easily understood example of assumption of the risk involves a scenario that, unfortunately, occurs somewhere in this country every summer. Virtually all people who have any knowledge of the game of baseball realize that foul balls are routinely batted into the spectator areas of baseball stadiums. Furthermore, those of us who are baseball fans realize that when we go to a baseball game and sit in certain portions of the stadium, we are exposing ourselves to the risk of injury from such a foul ball that intrudes into the spectator area.

Several court cases have involved unfortunate and serious injuries to spectators at baseball games who were struck by batted balls that landed in the stands. Almost universally, the courts reject these plaintiffs' cases and hold that it is common knowledge that, when an individual attends a baseball game and sits in those areas of the stands where there is a likelihood that batted balls will intrude, a particular patron assumes the risk of injury if she is so unfortunate as to be hit by a batted ball or thrown bat. As a further buttress to the voluntary exposure by the spectator to this danger, most courts take note of the fact that baseball stadiums routinely provide a screened area behind home plate where persons can select a seat if they wish to avoid the hazard presented by foul balls.

Inherent in the theory of assumption of the risk is the knowledge on the part of the plaintiff of that risk. In aviation cases, assumption-of-the-risk defenses usually fail. Most often this is because the plaintiff is able to show that he did not have an appreciation of the nature and extent of the risk presented to him, and therefore could not voluntarily assume it. To the extent that the baseball stadium cases have held assumption of the risk to be a successful defense, the courts have held that virtually all Americans have been exposed at one time or another to the rudiments of baseball and are aware of the danger of batted balls coming into the spectator areas. So it must be kept in mind that, in order for assumption of the risk to be a workable defense in a negligence case, the defense must be able to show that the plaintiff had a significant appreciation and knowledge of the danger, and then voluntarily exposed himself to it. Therefore, in aviation cases particularly—unless the plaintiff is another pilot or someone intimately familiar with the intricacies of flying—assumption of the risk is not a very effective defense.

Contributory and Comparative Negligence

In many cases, there is a significant likelihood that the plaintiff himself was not acting in a manner that would leave him totally free of blame. If the defense is able to show that the plaintiff was in any manner himself negligent in not performing the duties that he owes to others and to himself, and that failure was a contributory proximate cause of his harm, the defense of *contributory negligence* might become viable. In many states a few decades ago, contributory negligence was a defense that, if its elements were proved, was a complete bar to any recovery by the plaintiff, and the plaintiff went home from her trial with nothing. Under these rather onerous rules, there was no consideration as to the total quantity of negligence committed by both the plaintiff and the defendant, nor was there any weighing of the respective amounts of that total negligence committed by either the plaintiff or the defendant. In a state that recognizes a pure contributory negligence defense, if the defendant is successful in proving that the plaintiff was negligent himself in any manner, and that negligence was a contributing cause of the accident, the defense wins a total verdict and the plaintiff fails to recover anything. One nuance of the defense of contributory negligence is that it is axiomatic that a plaintiff owes a primary duty to himself to act in a manner that protects his life and well-being. Therefore, as a part of contributory negligence, if a plaintiff could be shown to have engaged in conduct that fell short of not only the duty that he owes to others but the duty that he owes to himself to protect himself, it was possible to exonerate the defendant from liability.

As tort cases became more common, legal writers, the courts, and lastly the legislatures began to see what was perceived to be the inherent unfairness of a pure contributory negligence defense in which a plaintiff could lose everything if his own negligence contributed only 1 percent toward the total negligence of all persons concerned in the case. An answer to that perceived unfairness has been the development of the doctrine of *comparative negligence*. Comparative negligence is a defense, but it is not complete in some circumstances. Under the doctrine of comparative negligence, the jury is told to compare the negligence of the plaintiff and that of the defendant by virtue of assessing percentage amounts of the total of all negligence attributable to the plaintiff and the defendant, respectively.

In some states, such as Ohio, there are cutoff amounts of negligence so that when the negligence attributable to the plaintiff reaches or exceeds those cutoffs, the plaintiff, then receives nothing, just as he did under the older pure contributory negligence doctrines. For instance, in Ohio, if a jury determines that the plaintiff bears 51 percent or more of the total of all negligence present in the case, the defendant wins and the plaintiff receives nothing. In situations where the plaintiff is assessed a lower percentage of negligence, the jury then determines the award that would be levied for that plaintiff if she bore no negligence herself, and then the jury verdict is simply reduced by the applicable percentage.

To further explain these workings, let us assume that the jury finds that the plaintiff suffered a total of $100,000 in damages from all elements, both special and general. If the jury then determines that the plaintiff is 30 percent at fault, the plaintiff would have her verdict reduced by 30 percent; in other words she would retain a $70,000 judgment against the defendant. Again, in a state with a comparative negligence law like that of Ohio, if the jury in the same situation found that the plaintiff was 51 percent negligent, she would receive nothing.

Some states have a comparative negligence law that allows the plaintiff to recover something, regardless of how high his share of the total negligence is, as long as it is not 100 percent. In these states, even if the plaintiff is found to be 70, 80, or 90 percent at fault, he still receives 30, 20, or 10 percent, respectively, of the total amount he would have received if he had been totally free of any negligence that contributed to the accident.

There is another move afoot in some states which seems to confound some lawyers and judges. Ohio recently merged the defenses of assumption of the risk into the defense of comparative negligence. The idea behind this theory is to allow the jury to determine if the plaintiff was somewhat negligent by assuming a known and appreciated risk.

It is left to the reader to decide which system is best. There are those who believe that once a plaintiff's negligence exceeds one half of the total negligence in the situation, he should receive nothing. That is the view taken by the law operative in Ohio. There are others who think that the plaintiff should still receive something if a defendant contributed at all toward the accident. Those are the operative laws in the states that exercise pure comparative negligence doctrines, allowing the plaintiff to receive only a small portion of what he otherwise would have received when he bears a large percentage of the total negligence that caused the particular accident.

Last Clear Chance

The doctrine of *last clear chance* is available to the defendant even though the plaintiff is able to show that he breached a duty owed to that plaintiff (and that breach of duty was the proximate cause of the accident), when the defendant is able to show that even though all those elements of negligence are operative in the case, the plaintiff had a last clear chance to avoid the accident and failed to avail himself of it. An example of a last-clear-chance defense is a situation where a motorist is driving left of center and continues driving her automobile on the left side of the road for some distance.

Let us assume that the highway is not heavily traveled at the moment, and the driver of an oncoming vehicle sees the wrongdoer left of center in more than enough time to slow his car and pull it over to the side of the road, allowing the person driving left of center to pass by, thereby avoiding a collision. If the plaintiff, the person who was driving on the correct side of the road, simply maintained his course and that resulted in a head-on collision, under the defense of last clear chance, the plaintiff can hardly be heard to complain. It was he who had, clearly, a chance to avoid the accident by simply pulling his car to the right, slowing it, and allowing the otherwise wrongdoer to pass by without collision.

In aviation cases this doctrine is obviously not applicable, except in very rare circumstances when passengers are injured in aviation accidents. It could be applicable to air taxi accidents, landing accidents, and similar situations when the pilot of an aircraft suffers an accident that is truly the fault of another, but which accident he had a last clear chance to avoid, had he chosen to take safe action to do so.

In summary, there are legal theories available to a defendant to avoid negligence liability, even though he has breached a duty to another that was a proximate cause of an accident injuring the plaintiff. This list is certainly not intended to be all-inclusive—there are other, less often used, affirmative defenses. The law does, in some circumstances, recognize excuses for conduct otherwise wrongful. But the best course of action for all us aviators is to know what we are supposed to do, and do it. Then, no one gets into court; but most important, we all fly more safely.

8

Particular Applications of Negligence Law

THE PRINCIPLES OF NEGLIGENCE LIABILITY OUTLINED IN CHAPTER 7 should make it clear that the general principles of negligence can apply to virtually any activity or endeavor in which one engages. Because *aviation* is such a broad term, and the aviation industry encompasses so many different activities, it is in order to take a look at how the principles of negligence law apply to a few of those doings. The segments chosen for a closer examination and discussion of negligence law, as it applies to them, revolve around the general aviation portion of this large industry. There is no discussion in this limited work of the law, including but not limited to negligence, as it applies to scheduled airlines or air taxi operators. Many of the same comments and explanations that are discussed in this chapter can, by reasonable analogy, often apply to those common carriers.

The areas of the general aviation industry that are to be discussed in more detail in the pages that follow are those to which most general aviation pilots are exposed at one point or another in their careers, or are business endeavors in which many general aviation operators do engage.

THOSE WHO RENT AIRCRAFT TO OTHERS

When an owner rents an aircraft to another pilot, for that other's operation and use, the person who does the renting (commonly known as the *lessor*) is exposed to liability if he is negligent. The negligence that the lessor can commit can take various forms, but primarily revolves around two areas. First, he has a *duty,* not only to the pilot to whom he rents the aircraft, but also to other people who are reasonably foreseeable to come into contact with it, to maintain the aircraft in an airworthy condition. In addition to that, the lessor has a duty to ascertain that the pilot to whom he is going to rent the aircraft is properly qualified and competent to fly it.

None of us with any good sense at all want to fly intentionally in unairworthy aircraft. It seems to go without saying that the lessor of an aircraft has the duty to keep it in airworthy condition for the safety of the pilot who is going to be flying it, the passengers, and anyone else who the law will say is reasonably foreseeable to come into contact with that aircraft. That is the standard of the duty of maintenance of the aircraft. This duty is owed to all persons who might reasonably foreseeably come into contact with the aircraft. But as we all know, there is maintenance, and then there is maintenance.

In the very recent past, the FAA has gone through a series of vacillations concerning what constitutes an airworthy aircraft in terms of what components, accessories, and what options on it can be nonfunctional and still allow the aircraft to be airworthy.

For a while, section 91.213 of the FARs required that any component of the aircraft must be operable, or completely removed, in order for the aircraft to be considered airworthy, unless there was a specific minimum equipment list generated by the operator of that aircraft, submitted to the FAA, and approved by it. This principle was well intentioned, but became ludicrous in its application, because, taken to the extreme, the requirement that everything on board the aircraft work was extended even to the cigarette lighter. We all know that there are certain redundant features of almost any airplane, particularly in the avionics, and a failure of one independent component or accessory does not render the aircraft unsafe in practical terms.

The FAA has recently made an about-face, and amended the regulations. Today, the basic requirement that everything on board the aircraft be working is still applicable, except that the decision is now left to the pilot who is flying a piston-powered airplane, and not flying in air commerce, to determine if any particular nonfunctional part is necessary for safe flight. If the pilot or operator makes the determination that one particular nonfunctioning item is not necessary or critical to safe flight, he has the obligation to do two things if he wants to fly the airplane before the item is repaired. First, he must placard the particular device as being inoperative, and then make an entry in the aircraft's logbook that such item is inoperable, and that he has made the determination that the inoperable character of that device is not critical to safe flight.

The lessor of an aircraft will undoubtedly have times when one or more such accessories, components, or other devices are inoperable, and that lessor will make the determination that such an item is not critical to flight safety. Be careful, and make sure that the proper placard is placed on or near the particular device not working, and also be certain that the required logbook entry is made.

A problem then arises in the event of an accident. (Hindsight is always 20/20, and foresight is never so good.) After an accident, years will go by when investigators, lawyers, insurance adjusters, expert witnesses, and finally, a court, will take multiple opportunities to second-guess a decision that was made prior to the mishap, and undoubtedly criticize it at length.

Let us take the example of a typical fuel exhaustion accident. Assume that the renter pilot got lost and was ambling about the sky and simply failed to find an airport and land before she ran out of fuel. Then expand the facts further and assume that the aircraft was equipped with two VOR navigation receivers as its means of electronic navigation; one receiver was nonfunctional. The lessor had made the determination that it was not necessary for safe flight, had properly placarded it as inoperative, and had made the required logbook entry.

Our pilot, who survived the accident with serious and disabling injuries, is now in court making the allegation that her ability to navigate was severely hampered by the inoperative character of one of the two VOR receivers and that is one of the main reasons that she got lost. She goes on to maintain that, had both receivers been operable, she could have done a far more professional job of navigation, could have cross-checked her position with far greater ease, and naturally, would not have become lost, and therefore would not have run out of gas, and would not have suffered the accident in question.

Now put yourself in the place of a lay juror sitting in the jury box, with no concept in the world about anything having to do with flying airplanes. That juror might think that airplanes are dangerous in the first place, particularly light aircraft, and to rent one with an inoperable navigation receiver is foolhardy. So it is easy to see that, although a regulatory minimum might be met and complied with, purely meeting that regulation will not be a saving event in the case of an accident and a subsequent lawsuit.

Every lessor of rental aircraft must realize that any time he puts an airplane on the line that is not in absolutely 100 percent tip-top condition, he is exposing himself to criticism and potential liability if a jury finds that he breached his duty to supply a properly maintained and airworthy aircraft to those who rent planes from him. I could go on at length with numerous examples, but I believe that the preceding one is sufficient, by analogy, to cover virtually any inoperable part or component of the airplane and the way that that situation might well be viewed after an accident.

If you are in the business of renting aircraft to others, take heed. The prudent, cautious lessor will strive to keep his rental fleet in good condition, and promptly repair even those items that, in the view of an experienced pilot, might not be critical to flight safety. In addition to the legal reasons for so doing, it makes good business sense in a competitive environment. Most renter pilots appreciate flying good equipment, with avionics and other accessories that work when they are needed, and like to fly aircraft that are clean and exude a feeling of well-being and confidence.

One of the first cases in which I was involved as a lawyer revolved around a potentially lethal problem with a rental airplane. Fortunately for all concerned, no one received any personal injuries.

On a cold, winter morning, a student pilot and his flight instructor went out to fly in a Cessna 150. A normal preflight inspection was conducted, and both of them got into the airplane. Because it was cold that early morning, and the 150's battery was apparently weak, the little Continental engine would not start.

The instructor had the airplane towed into the flight school's hangar, where a line service person opened the cowling for the purpose of changing the battery. When he began to take the cover off the small metal box that contains the battery, he noticed a white, gummy substance around the battery box and oil filler neck.

Thankfully, the service man called a mechanic over to have a look. The mechanic became suspicious, as the material was obviously foreign to any proper aircraft application in that particular area. Because the mechanic's suspicions were quite strong, he wanted to take a sample of this material for further inspection. He scraped some of it off the battery box and went across the airport property to seek the opinion of some experienced individuals at a well-known and well-reputed engine overhaul shop. Before too much time elapsed, the individuals at that overhaul shop, together with the mechanic employed by the flight school, all came to the fairly certain conclusion that the substance was some abrasive, probably valve grinding compound.

It then became apparent what had happened. Someone had attempted to sabotage the airplane by introducing valve grinding compound, which is extremely abrasive, into the engine's oil supply. Obviously, the FAA got involved, as well as the law enforcement authorities.

The upshot of the case as far as the lawyers were concerned was simply to arrange—on behalf of the aircraft owner's insurance company—a complete teardown, inspection, and major overhaul of the engine. Even though it appeared that the engine had run, if at all, for only a short period of time with the abrasive in its oil supply, the engine was extensively damaged, and virtually ruined.

Had the battery not been weak, had the morning not been so cold, or had the flight been scheduled for later in the day when the temperature would have been higher, the end of the story might well have been quite different. Most of the individuals involved in the inspection of this engine, who are certainly more knowledgeable about those things than am I, estimated that the engine would probably have run no more than 30 minutes to an hour until it completely destroyed itself, and either suffered a catastrophic failure, or at least quit running.

This possibly tragic story has a happy ending, in that no one flew the airplane before the attempted sabotage was discovered, and the owner of the airplane received a newly overhauled engine, courtesy of his insurance carrier. The bad part of the ending is that the law enforcement authorities, as far as I know, never were able to discover who the saboteur was.

The point is quite simple—keep a sharp lookout for the mechanical condition of the airplanes in any flight school's rental fleet, or any other lessor's operation. Stranger stories than this have occurred, and sick minds often seek to wreak their vengeance through destruction of not only vehicles, but lives as well.

In addition to the foregoing comments concerning the mechanical condition of the airplane, the lessor has a very definite duty to ascertain the pilot qualification and competency of his rental customer. Thankfully, the requirements of insurance companies today demand, in order for rental coverage to be effective and protect the lessor's interest, that pilots go through a process that is generally designed to ascertain their competence.

First of all, no lessor should ever even think of renting an airplane to a pilot without first examining three things. The first is the pilot certificate possessed by the potential renter. Next his medical certificate should be viewed and particular attention paid not only to the class of it, but the date upon which it was issued, so that a quick mental calculation can be made to determine if it is still valid. Last, the lessor should examine the pilot's log to compare it to whatever statements the pilot might have made concerning his total flight time, time in this particular type of aircraft, date of biennial flight review, and anything else he might have said about his flight experience.

Have the renter pilot fill out an application to rent the aircraft, and make that application detailed. Ask questions about his total time, his recent experience, the date of his most recent flight review, the types of aircraft he has flown, the areas of the country into which he has operated, the amount of instrument time that he has, and similar areas. Then compare the answers on the application with his log.

Assuming that the potential renter passes this scrutiny, a complete and thorough flight check is in order. Simply complying with the insurance requirements is not enough. The flight check needs to be tailored to the particular renter in question. If the individual does not fly very frequently, if he is a low-time pilot, or if he has not flown this specific type aircraft before or in the relatively recent past, the initial checkride needs to be extremely thorough and all-encompassing.

The renter pilot must be made to study the pilot operating handbook and should be quizzed at length about the critical performance characteristics of the aircraft and the emergency procedures delineated in the handbook. The check pilot should keep notes of this quizzing, and better yet, the lessor should devise a short written quiz to see if the customer has actually reviewed the manual in-depth and has an understanding of it.

Perhaps the renter pilot needs a biennial flight review (BFR) at the same time that he is undergoing the qualification check to determine if he can rent the aircraft. In the examination of the pilot's log, be absolutely certain that he is within the time limits since his most recent flight review. If you have any doubts about the potential renter's currency, or knowledge of the general operating procedures outlined in Part 91 of the FARs, a wise FBO or other rental operation might insist on a BFR before renting the aircraft to the customer.

The old days of two or three touch-and-goes comprising a checkride for rental purposes are gone in most good operations, and should be relegated to nostalgia everywhere. An initial check of a renter pilot should be very thorough, and should basically include all the maneuvers that would be required in a biennial flight review.

Unfortunately, economics sometimes force aircraft lessors to shorten the qualification process because customers complain about the amount of money spent for aircraft and flight instructor during this process. Perhaps a few potential renters will not be willing to subject themselves to a complete check, and if they are reticent to do so, my personal belief is, so much better for the operator if they are eliminated from the rental ranks. If the qualification process is used as a period of instruction and so presented to the renter pilot and then carried out as continuing education and recurrent training, pilots with the most elementary notion of safety will not object.

The law of negligence recognizes the concept of *negligent entrustment* of a dangerous instrumentality to one not properly qualified to use it. Although aircraft are not generally defined as dangerous instrumentalities, as the law defines such items as firearms, archery equipment, and similar devices that are designed to render physical harm to something or someone, the concept of negligent entrustment can easily work its way into a case that results from an accident involving a pilot whose qualification or competency is at issue. Remember that plaintiffs are always looking for some source of recovery.

The more defendants who can be brought into a case, and against whom a maintainable legal theory can be sustained, the more parties there are exposed to eventual liability for the damages suffered by the person bringing the suit. In such a situation, more than one insurance company will likely be in the case defending and will want to, if the circumstances justify it, discuss settlement with the plaintiff for sums considered reasonable by that insurer. Most plaintiffs' lawyers are always looking for some deep pocket that they can invade to accomplish their goals of compensation for their clients.

Merely because the pilot might be at fault in the accident does not keep the lessor from being in the case with her if it can be shown that the lessor was not completely thorough and otherwise proper in his attempts to determine the pilot's qualification and competency prior to the rental of the aircraft. In fact, it might be not only an injured passenger but even the incompetent pilot herself who gets injured in the accident, and then goes against the lessor of the aircraft, claiming that that lessor had no business renting the aircraft to her in the first place.

There have been suits filed, and settled for large amounts of money, by pilots who have suffered serious injury, and sometimes death, in accidents where they, or their estates, then claim in later litigation that they were not qualified to fly the airplane in question, and the lessor knew it and should have exercised the duty to recognize that fact, and not rent the airplane to that pilot.

This same attack may be made against the lessor by passengers in the aircraft or by anyone else who the law will say should have been reasonably foreseen by the lessor to come into contact with the aircraft. That could be people injured on the ground, or those who own property that was damaged as a result of the accident as well.

I recently defended an FBO in just such a case. A Cessna 172 had been rented to a customer who announced his intention, after checking it out with the FBO, of making a cross-country flight of about 150 miles to the destination, and returning that same day. The landing lights on the airplane were not functional, and the panel was properly placarded to reflect the discrepancy. The customer began his flight relatively late in the day, when it might have been possible for him to complete his round-robin flight before dark, but at a time when it was very likely that he would be returning well after sunset.

He did just that—he got back well after dark, in fact, several hours after sundown. His landing was uneventful, but as he was taxiing to the FBO's ramp, he collided with the tail section of a Cessna 182 properly parked and tied down in another ramp area, which was not lighted. The renter pilot never saw the 182 until he heard the sickening sound of ripping metal. The owner of the 182 sued the pilot and the FBO, claiming that it was negligent for the FBO to rent the airplane to this pilot for this trip, beginning at the particular time of day. As luck would have it, the owner of the 182 had purchased the airplane new from the factory, and showered it with loving care for all the years she owned it.

The airplane was repaired. However, since what had been a pristine 182 now had a history in its logs of major damage, its market value had been diminished, even though it had been very competently repaired. The FBO's insurance company paid the lady some significant money.

Lessors should also require a periodic checkride of the regular, long-term renters. Many rental operators allow pilots to fly without recheck if they fly at that facility with some stated frequency and interval. Such a pilot could be just as dangerous as one who was never able to pass an initial checkout. The essence of recurrent training recognizes that we all lose what we do not regularly practice.

In the last analysis, be careful, and do everything possible to rent airplanes only to pilots who are qualified, current, and properly tested in the particular type of aircraft.

FLIGHT INSTRUCTORS

The certificated flight instructor is quite probably the most important influence upon the later career, safety, and attitude of any pilot. I doubt seriously that there are any of us who do not have very distinct memories of our first primary flight instructor, and hopefully the memories are favorable and pleasantly reminiscent of a time in our lives when we were being introduced to an entirely new environment. Unfortunately, I know some pilots who do not have the same fondness toward their initial training as do the fortunate people who look back on it with warm recollection.

The flight instruction profession is an enigma in its own right. It is most often populated by relatively young and low-time pilots serving what they—along with a good part of the industry—feel is a period of time in the trenches, building pilot-in-command time until better opportunities open in a corporate flight department, an air taxi operation, a regional airline, or, the highest goal of most young flight instructors, a seat in the front of a regularly scheduled airliner. I was no exception to this general state of affairs. Exactly 1 year and 1 week after my first solo flight, I was occupying the rear seat of an Aeronca Champ at the ripe old age of 18, giving flight instruction to people two or three times my age.

Those were simpler times. The airplanes were about the simplest pieces of machinery ever to take to the sky since the 1930s, the airspace was uncomplicated, and for the greatest part, uncontrolled. We flew hundreds of miles across the country in aircraft without electrical systems and, necessarily therefore, without any kind of electronic navigation equipment or communication radios.

Today only a very few students begin their training in that type of aircraft. While it is not the purview of this work to engage in a discussion of personal beliefs concerning the state of affairs in modern flight training, it is enough to say that many of us believe that, if the economics of the situation allowed, many pilots would benefit from being taught basic aviation skills in simple airplanes before they progressed to more complex equipment and a more complicated environment. However, there are many good arguments to the contrary, and this is a controversy to which there is no absolutely correct answer.

Flight instructors bear a significant liability, not just during the actual giving of flight instruction, but often continuing long after. During an actual training flight, most particularly in the stages of a student's instruction before she gains a private pilot certificate, the instructor is, without doubt, the pilot in command of the airplane during all phases of any dual instruction flight.

Hence, the instructor has all the liability, not only to the student, but also to any other person who might reasonably be foreseen to come into contact with that flight. If an accident arises, it is the flight instructor who will answer to the FAA and the civil courts as well. In the precertificate stages of a pilot's training, he depends completely upon his instructor for the safe conduct and termination of each dual flight. In fact, because the flight instructor is deemed pilot in command of the aircraft, a student has absolutely no authority to contravene any instruction given him, or any decision made by the flight instructor.

I have clearly established the fact that the instructor has the duty to conduct the period of dual instruction safely. If he breaches that duty and an accident occurs as a result of that breach of duty, the instructor can expect to be personally named in any lawsuit that emanates from that accident. As well, others might be made defendants to the same suit, but that discussion follows.

A pilot more often than not develops and retains his basic attitude toward piloting from the influence of his first flight instructor. The instructor has a duty not only to fly the airplane safely while he is giving instruction to his students, but also to instill—in addition to knowledge—basic concepts of safety, judgment, and attitude in those same trainees. Any experienced pilot knows that flying is not just a mechanical skill. There are many competent and safe pilots whose physical ability to manipulate the controls, while meeting the certification standards set forth for their certificate, are not what some might consider to be superb. But as with most other things in life, piloting skill is a compromise of many subskills. Judgment and attitude can go a long way toward the making of a safe pilot, even if her manipulative skills are, while standardized, not as high as she would like them to be.

To the converse, there are, unfortunately, a few folks in our industry who, from a physical point of view, have been said to be able to fly the crates the airplanes come in. But their judgment and attitude are so reckless that they are nothing but an accident waiting to happen. Whose fault is that? Did the original flight instructor who began these folks' careers in aviation do his job, or did he simply teach them to mechanically control a machine, without understanding that such is only a part of the development of a safe pilot?

Flight instruction takes place under two similar but different circumstances. Most flight schools operate as FAA-approved training centers, approved under Part 141 of the FARs to conduct flight instruction pursuant to a specific curriculum for various certificates and ratings. Other, usually smaller, operations and freelance instructors operate under Part 61 of the FARs. In such cases, there is no FAA approval to their schooling, other than that which was inherent in their becoming flight instructors in the first place, and there is no FAA-approved curriculum for their teaching aside from that set forth by analogy in the flight test guides and portions of Part 61 that determine the necessary training that students must receive when seeking certain levels of certificates or ratings.

Part 141 requires not only an approved curriculum, but also a great deal more paperwork on the part of the flight school. An essential part of the documentation required under Part 141 is a student progress sheet, tailored to fit each certificate or rating for which the school is approved. The student progress sheet may have another title or name attached to it, but its main purpose is to provide a permanent record to ascertain that the student has received instruction in each and every element set forth in the curriculum.

This progress sheet usually calls for the date that each lesson was given, a place for the flight instructor to certify that he gave it, and perhaps an area for comment by the flight instructor concerning the conditions of the dual flight and the performance of the student as well. Further, Part 141 schools are required to provide structured ground school for the ground training required to achieve each particular certificate or rating, and to thoroughly document that as well. Approved flight schools also generally have the requirement to conduct two or more stage checks throughout the student's training, so that another flight instructor can fly with a student to give him the opportunity to evaluate that student and, as an indirect result, see what kind and quality of instruction the student has received.

If named in a lawsuit, a Part 141 school or its instructor, if regulations have been followed, will have a very detailed and thorough documentation of a student's training, progress, performance during training, and outside evaluations of that progress at various stages during the course of instruction. Unfortunately, there is no such requirement for a school not certified under Part 141, nor for the free-lance instructor.

When a training operation, including but not limited to, a free-lance instructor, is not approved under Part 141, it must then conduct its endeavors pursuant to Part 61. Part 61 is that part of the FARs that sets forth the aeronautical skill and experience requirements for trainees to achieve the various certificates and ratings that are thoroughly detailed in Chapter 2. There is no requirement in Part 61 for record keeping, other than the logging of training flights and solo flights in the student's and instructor's personal pilot logs. Beyond that, each instructor in a Part 61 operation is left to his own resources and standards of training and record keeping.

As this edition of this book is being written, a proposal is being circulated by the FAA to significantly overhaul Part 61 of the FAR. If adopted in the form in which the changes are proposed, one of the new mandates will be for Part 61 instruction to be conducted pursuant to a written syllabus. If you are a flight instructor, be on the watch for Part 61 to receive an update, and make sure that you are aware of the new requirements, whatever they end up being.

There can never be too much said about the necessity and benefit of complete and thorough training records. Most of the time when an instructor gets sued, the plaintiff makes an allegation that the instructor failed to properly teach, or completely omitted instruction in, particular areas that good practice would dictate be taught. Assuming that a particular suit involves an accident that occurs during a student's solo time, or perhaps later, as will be discussed in a short while, the flight instructor who has not done anything beyond the logging of the elementary rudiments in the student's personal log and his own is at a distinct disadvantage.

How can a flight instructor show that he gave a student a thorough lecture and discussion of density altitude effects upon the performance of an airplane? If he operates under Part 61 rules, this instructor had no obligation to record that ground training time. If he did not record it, all he can do is sit on the witness stand in a courtroom and say that he did it, but without any backup to substantiate his testimony. Juries and courts realize that a defendant's recollection of events relevant to the lawsuit tends to be beneficial to his position. If not, there would be no case in the first place.

Free-lance instructors and instructors operating in Part 61 schools should absolutely keep records with the same detail that is required of Part 141 approved flight schools. The Part 61 instructor should devise a ground training program that is as thorough as possible, and further create a progress sheet, so that he can not only log flight time and flight training in the student's pilot log, but also keep permanent record of all discussion time, ground school, and tutoring that occurs during the process of making a raw student into a recreational or private pilot. It is also good policy to have the student initial or sign the training entries along with the flight instructor. That prevents any later allegation that the records were concocted after the fact in an attempt to create a defense.

After a student completes her training and gains the certificate or rating for which she came to the instructor, the instructor's liability is not at an end. There have been many cases filed, settled, or tried in which a pilot had an accident some significant period of time after the completion of training, and either the pilot or a passenger makes a claim against the flight instructor. It has been claimed that an instructor either poorly trained the pilot in a particular phase of operation, or that such training was omitted altogether, and an accident resulted.

There is generally no defense for omitted training. There will always be the incompetent among us, and if a flight instructor chooses to either forget or intentionally delete training that should, in the proper practice of his profession, be given to a student, he will be at a tremendous disadvantage trying to defend his conduct.

Training that is alleged to have been poorly conducted presents a slightly different problem. If the student survives the accident, even though he might now be a certificated pilot, and if the instructor documented the training sessions that were given, the situation might degenerate into a contest in the courtroom between the student's story and the instructor's. The student might maintain that even though instruction was given, he did not thoroughly understand it, that the instructor knew that he didn't understand it, and this lack was glossed over by the flight instructor and training went on without one of the building blocks securely in place. There is no easy answer to that problem. Instructors can only do their best. Defense lawyers cannot totally eliminate exposure, and thorough practices on the part of the instructor will almost always serve to reduce his risk, but it can never be totally ignored or forgotten.

There was a case some years ago where a fairly newly certificated private pilot put four people in a Piper Cherokee 140. He proceeded to take off from a relatively short runway, but the same runway from which he had flown since the beginning of his training. Unfortunately for all concerned, the new pilot had learned to fly during the winter season, and this was one of his first exposures to flying in warmer temperatures. The airplane performed as expected in a high-density altitude environment. It did not clear obstacles at the end of the runway, obstacles it had cleared many times in the past for this pilot during his training, when the air was crisp and cool, and the airplane performed accordingly.

Several people, both in the airplane and in the building into which it crashed, met their demise. It does not take the village genius to make an educated guess at who was one of the target defendants in a rather massive lawsuit. The flight instructor who trained the pilot killed in the accident was alleged to have not properly taught this individual the effects of density altitude upon the performance of this airplane, particularly when it was operated at or near its maximum gross takeoff weight.

The result was not good for any concerned. A pilot, his passengers, and totally innocent individuals on the ground were killed in an accident that should never have happened. I do not have specific knowledge about the actual training that this pilot received, but for some reason he chose to attempt a takeoff that was doomed from the moment he first advanced the throttle at the end of the runway.

Another area in which flight instructors suffer potential exposure to negligence suits is the providing of biennial flight reviews to pilots who are already certificated. All professional pilots have some form of recurrent training required by their employers or insurance company. Corporate flight departments routinely send their pilots outside their own operations for very highly structured and formalized training, usually at yearly intervals. Air taxi and airline operators have recurrent training requirements contained within their operations manuals, and for them it is a violation of the FARs to breach mandatory duties contained within those manuals.

The general aviation pilot who is not engaged in commercial operations has virtually no required recurrent training, other than the biennial flight review that must be administered every 24 calendar months.

When the biennial flight review came into existence in the early 1970s, it was met with a significant amount of resistance from the aviation community. Most of this complaining was unfounded, and time has shown that the process is not particularly expensive or burdensome, and has produced many positive results.

So for most private pilots and those who hold higher certificates but exercise only private privileges, the sole recurrent training to which they must expose themselves is the biennial flight review. Flight instructors need to keep this sorry state of affairs in mind when administering reviews; for it is only during the conduct of a flight review that whatever weaknesses the particular pilot might have in judgment, attitude, and flying skill can come to light.

The flight review is set up in the regulations so that it is not a pass-fail procedure. Because the review must be completed by every pilot within the abovementioned calendar time restrictions (biennially), the flight instructor who gives it is presented with the only opportunity that this pilot might have during that 2-year period of time to get any meaningful instruction and corrective activity.

The regulations leave the conduct of the review up to the instructor giving it, with the only regulatory requirement being that the review must have a minimum of 1 hour ground discussion and 1 hour flight time in powered aircraft. For glider pilots, the rule is still 1 hour of ground instruction, and a minimum of three glider flights. There is an advisory circular available from the FAA for guidance: *Scope and Content of Biennial Flight Reviews*. A perusal of FAR sections 61.57(a) and (b) shows that these are skeletal sections at best. One of the perceived deficiencies in the flight review system, about which there is significant discussion toward change at this point, is the fact that a certificated pilot can accomplish the flight review in any aircraft for which she is rated. For instance, I possess a commercial pilot certificate, rated for single-engine airplanes, multiengine airplanes, rotorcraft/helicopters, and gliders. I can, under the bare requirements of the regulations, accomplish a biennial flight review in a two-place training glider, and that review is sufficient for any aircraft which I may choose to fly, from that same glider to a Beech King Air, to a Bell JetRanger helicopter.

Do not doubt that there are some pilots who feel that the biennial flight review is an imposition upon their time and pocketbook, and questions their skill as pilots. Long ago, a police officer once told me that he had found, in the process of citing hundreds of miscreant drivers, that very few people will ever admit they are either a bad lover or a bad driver. In our industry, I would extend that comment to pilots—I have not found one yet who would admit that he suffers in his flying skills.

Someday the regulations might be changed to require that each pilot who wishes to exercise all the privileges of his particular certificate accomplish reviews in each aircraft for which he is rated, or at least, through some determination or another, the most complicated or difficult to operate of the aircraft for which he is rated. But because that is not presently the case, the foolhardy are permitted to continue on in their ways if they so choose, and train on a recurrent basis in the simplest aircraft for which they are rated, and use that flight review to suffice in the most complex.

When a flight instructor gives a flight review to a pilot who is rated for more than one category or class of aircraft, the instructor has no choice but to proceed with the review in the aircraft chosen or presented for its conduct; unless he simply does not want to fly with this individual in the first instance. But assuming that he does, because that is how he makes his living, the instructor should conduct the review carefully, and then note in the pilot's log that the review was conducted in that particular type aircraft. Do not leave it to the pilot being reviewed to make the log entry. Also, I would recommend that instructors, when using the language in FAR sections 61.57 (a) and (b), expand beyond the required minimum terminology and specifically set forth in the log entry that the review was conducted in a specific airplane.

In the last analysis, no flight review can be too complete. When I was an active flight instructor, I would not perform a flight review for a pilot possessing an instrument rating—even though he might not meet the currency requirements of Part 61—unless it was coupled with an instrument competency check. A court or jury might never be able to thoroughly understand the practical side of how flight reviews are often viewed by instructors, and those being reviewed. If an instrument-rated pilot performs only a VFR flight review, and then goes out and suffers misfortune during an instrument flight, the flight instructor who administered that VFR-only review could be the target of some serious scrutiny, including, but not limited to, being named a defendant in any lawsuit that results from that accident.

Some flight instructors become designated pilot examiners after invitation by their local FSDO. The system of pilot certification would not work at all if it were not for the corps of flight instructors who have been designated to serve as pilot examiners. Each pilot examiner is rated to give flight tests for only certain pilot certificates and aircraft ratings. Not all designated examiners are permitted to give all the flight tests; in fact, I do not know of any designated examiner who is permitted to give flight tests for all the imaginable combinations of certificates and ratings.

In the litigious society in which we live, even designated examiners must be aware that they too possess a certain degree of liability if they are negligent in the conduct of their duties. The law does realize that it is these examiners who stand as the final guard against the ill-trained or incompetent person becoming a certificated pilot.

The example cited earlier in this chapter, about the poor soul who tried to take-off in the Cherokee 140 with four persons on board on a hot summer day, did not end with the instructor being sued. In this case the designated examiner who administered the flight test to the deceased pilot was also named a defendant in the lawsuit. Unfortunately, he too was completely uninsured and was cast adrift in the legal system not only to pay for his own defense but also to face potential financial ruin, should he be found liable by the jury at the conclusion of the trial of the case.

The increasing exposure faced by designated flight examiners is relatively recent in the law, but is something that cannot be ignored and the exposure will probably go on to expand in the future, along with other areas of liability.

In summary, flight instructors have one of the most important and consequential functions within the aviation industry. They see to the care and feeding of new students, and to the continued proficiency of Part 91 operators who have no other requirement for recurrent training. There can be very little overly thorough instruction given, and in the process, all instructors should learn to create appropriate progress sheets and other documentation of all instruction that they give, both on the ground and in flight, and diligently maintain those records in perpetuity.

FUEL SUPPLIERS

Engineers will tell us that an internal combustion engine burns air, and needs only enough fuel introduced into the air to make the combined charge combustible. Those of us who are not engineers tend to think that an airplane runs on fuel. That is a fact of which I am still a firm believer, and many flights have come to grief either through the exhaustion of an airplane's fuel supply or occasionally because the incorrect fuel was put on board before the flight began.

The liability faced by fuel suppliers can be a combination of negligence liability and product liability. We will discuss product liability in greater detail in Chapter 9, and anyone who supplies fuel needs to have a thorough understanding of those principles as well as negligence.

Fuel can become unsuitable through many means. It can be improperly refined, it can become contaminated through improper transportation or storage, or it can simply be put in the wrong airplane. The refining process is a mystery to me, someone who cannot even begin to discuss how crude oil from Saudi Arabia becomes 100 low-lead aviation fuel in the tanks of my airplane. The process is mysterious but, thankfully, is one that so far has yet to fail me. If it does, the fuel supplier probably has a product liability exposure that can include concepts of negligence liability, but the other theories set forth in Chapter 9 might be operable as well.

After the fuel is refined, it must go through several stages before it ends up in an airplane's tanks. It will probably be stored for some period of time at the refining site, then transported to a distribution center where it might undergo another period of storage, and then transported again to the airport or fixed-base operator which will pay the supplying company for it and take over the care of that fuel from that point to the aircraft fuel tank.

During each of these stages, the fuel supplier has the obligation to use the appropriate degree of care, which is high, to maintain not only the proper manufacture of the fuel but also the purity of it. Fuel contamination can develop from many sources. Fuel can be accidentally mixed with other types of petroleum products, or it can be transported or stored in a manner such that it comes into contact with contaminating elements—chemicals or products that render it unsuitable for use in aviation applications.

One of the major nemeses to aviation fuel is contamination with water. Water is heavier than gasoline and any condensation that occurs on the interior of a storage tank or fuel truck will settle to the bottom of the fuel load. Because most fuel trucks discharge their cargo through valves and discharge hoses that come out of the bottom of the tanker truck, any water contamination that has been introduced into the storage tank or truck, or that has developed in it during the time the fuel was contained therein, stands a very good chance of going with the fuel load into the storage tanks at the airport or fixed-base operator. The fuel supplier has a very high standard of care to take the appropriate steps to prevent water contamination, and in the final analysis, to discover it if it is present, before the fuel is placed in aviation use.

Fixed-base operators are also fuel suppliers, and a discussion of their potential errors follows in this chapter. Here we are concerned with the duties and potential breach of those duties that apply to the fuel supplier before the fuel gets to the fixed-base operator.

If an aircraft suffers an accident because of fuel contamination or another such problem, there is little doubt that competent plaintiff's counsel will seek to join the fuel supplier as a defendant to the suit. There have also been cases that have held the fuel supplier to be in a position of principal, in a principal-agent relationship, with the fixed-base operator that actually sells the fuel to the pilot or aircraft owner.

Those same cases have held that the fuel supplier can be liable, under a principle of law wherein the principal is *vicariously liable* for the negligence of his agent.

Most refining companies certainly do not intend to be in a position that they are answerable in court for misfueling errors and other such negligence on the part of fixed-base operators to which they supply fuel, but some recent court cases have put them in that position.

Again, it is not my intent here to get into a debate or serious discussion as to the propriety of the rules of law, for such contemplations are better left to those who can do something about it—judges and legislators. All I can do is make an attempt to inform those in our industry about the various areas and theories of the law that affect them and their operations. Whether it is eventually considered fair that a fuel supplier should bear ultimate vicarious liability for a fixed-based operator's fuel truck driver who puts jet fuel into a piston engine airplane is a question that cannot be answered here, and can probably never be given any final answer appropriate for all circumstances. Nevertheless, that liability has been held to exist, and fuel suppliers need to be aware of it.

MAINTENANCE OPERATORS

Today's modern aircraft are works of mechanical wonder. The combination of the propulsion system, the airframe, and all the associated subsystems, and the electronics and avionics in modern general aviation aircraft would probably mystify the aircraft mechanic of only a few decades ago. Every time something in life increases in complexity, the opportunity to err concomitantly increases with it. The maintenance of aircraft is no exception.

In similar fashion to flight schools, maintenance facilities can operate under two different philosophies. A maintenance facility can become an *approved repair station,* whereby it submits maintenance manuals to the FAA for approval and maintains the necessary tools, testing equipment, and personnel to meet a rigorously defined standard of operation. Many smaller maintenance shops do not seek FAA approval as an approved repair station, but rather operate under the *authority* of the particular mechanic certification possessed by either the owner or employees of that facility.

The approved repair stations most often are the ones that tend to maintain the upper end of the general aviation fleet, while a good portion of the single- and light twin-engine variety of airplanes are maintained, often just as competently, by mechanics who either own or are employed in shops that have not sought and received formal approval as approved repair stations. Some of the testing equipment that the regulations require for designation as an approved repair station would simply not be applicable or necessary in the course of business of many of the maintenance shops around the country. It is not the intent here to make any qualitative statement about whether a maintenance facility needs to be an approved repair station or whether it can operate just as well and as competently on an unapproved basis using the authority granted by the certificates held by the mechanics therein as authorizing the repairs and maintenance conducted.

When a maintenance shop does its business, it exposes itself to liability for three basic areas of its endeavors. First, it can be liable *for the improper conduct of repair and overhaul services* that it makes, including a vicarious liability for the negligence of subcontractors to whom it might send certain systems or assemblies for inspection or repair outside of the particular shop's own facility.

It is possible that a maintainer can also become liable *for parts that he supplies and incorporates into the repairs or overhaul services* that he provides. He has better defenses available here than in any of the other two areas, if he purchases parts that are properly approved from manufacturers of them. Again, the product liability discussion in Chapter 9 explains that manufacturers primarily bear this liability, and if the parts are purchased only from them and without further change or modification are passed on to the consumer, the intermediate maintenance facility might escape liability for the defective part. But many parts are fabricated in maintenance shops and then the shop has no one to whom it can turn to pass the liability along, if that part is defectively made. However, even parts made properly inside or outside the shop always have the potential for negligent installation.

The last main area in which maintenance facilities are exposed to negligence liability is *in their inspection function.* A problem cannot be repaired, properly or negligently, if it is not discovered during inspection. Probably as many cases are filed against mechanics and their employers alleging improper inspection and the following failure to perform a needed repair as are brought for the negligent conduct of repairs themselves.

Let us first discuss the act of performing repairs, and where negligence liability can result. Courts and juries are not particularly akin to feeling that aircraft repair, like many other aspects of aviation, allows much room for error. When an aircraft is brought into the shop for maintenance, the owner or pilot who bought it there is justified in relying upon that facility to use a very high standard of care in the performance of those repairs. Even though the law requires that the shop use that standard of care that is prevalent in the area for maintenance facilities, that prevalent standard is almost always a high one indeed. Most often a pilot will bring his trusty steed to the mechanic with only a vague description of something going awry. It is then up to the mechanic to take that complaint on the part of the pilot and reduce it to a problem that is ultimately properly repaired, and should not affect the safe conduct of the next flight.

The certificated airframe and power plant mechanic has one of the most arduous jobs in our industry. He is expected to be part clairvoyant, part engineer, and always answerable for the functioning of all aircraft submitted to his care. When that aircraft fails to function properly, it is that mechanic, or his employer, who is going to be required to respond as to why.

Sloppy maintenance has no business in our industry. If a pilot or aircraft owner brings an airplane into a shop with instructions to skimp on the conduct of the necessary repairs, my general advice to that shop is to tell the pilot to go elsewhere. If the pilot is exercising that kind of judgment in the maintenance of his airplane, the various other aspects of his attitude and judgment are probably equally suspect. He, again, is an accident waiting to happen. The prudent maintenance shop does not want to be a signatory to the logs of that airplane when it has its eventual accident.

Maintenance is expensive, and all of us who own an airplane are sorrowfully aware of that fact. Nonetheless, it is a fact of life for the aircraft owner. If a particular owner either does not want or cannot afford proper repairs that are thoroughly done, I would, if I were a mechanic, definitely shy away from working on that airplane. Even though good maintenance is expensive, most pilots, including myself, appreciate the sense of security that results from retrieving an airplane from a good maintenance facility, knowing we can rely upon the competence of the mechanics and the thoroughness of their work. Five thousand feet above the ground on a moonless night is no time to discover that the mechanic either was sloppy or deferred needed repairs to the economic pressures put upon him by the pilot.

If a maintenance shop is going to allow itself to be influenced to lower the standards of care that it uses in the performance of its work, that shop will eventually end up a defendant in a courtroom. Merely because the pilot does not want thorough repair work done, does not mean that his passengers or those on the ground would expect anything but compliance with a standard of care that the law requires of maintenance facilities.

Virtually all repairs and routine maintenance require that some part or parts be supplied by the shop, in addition to the labor necessary to install them. With the modern concept of design and manufacture of things as complicated as airplanes, providing systems in modules that are easily replaced rather than repaired—as was done in times gone by—the sale and installation of parts has become a major portion of maintenance activities. The normal maintenance shop's exposure to liability in this area comes from the *installation of the parts,* which must be done in a proper manner, and from parts that it fabricates.

Even though modern airplanes are extremely complex, they still have a good number of parts that are susceptible to local fabrication when replacement is necessary. Exhaust stacks are often welded, and certain hinges, fasteners, and other such devices are routinely fabricated in maintenance shops. The law requires that the same high standard of care that applies to the labor activities be utilized and complied with in the fabrication of locally made parts.

If a mechanic seeks to repair a part rather than replace it, he has the duty to do so properly. If he elects to weld a cracked exhaust stack rather than replace it, the repair needs to bring the part up to the same standards that a new part would provide. If he reworks a fastener on a baggage door, cowling, or door, likewise after his repair, that fastener is expected to perform its function to the same degree that it would if it were a new part properly installed. For this reason, among others that are purely economic, every day more and more maintenance facilities get out of the business of repairing parts, and concentrate more on pure replacement.

With the current legal environment, no one can particularly fault them for that attitude, especially when it is realized that, as with all other businesses, aircraft shops are in existence to earn a profit. The markup on the sale of new parts often exceeds the potential for profit that can be gained from the expenditure of labor to repair a part that is broken. As well, the new part will likely last longer and give better service than one that only has a spot repair administered to remedy an immediate and narrow problem.

Lastly, and very importantly, maintenance facilities can suffer liability for *improperly performing their inspection function.* Airplanes need repair when things go wrong. But even what appears to be a perfectly functional airplane is required to undergo periodic inspections, and for the operators who fly light aircraft under Part 91 of the FARs, this periodic inspection is mandated annually.

A great deal of fear and trepidation often accompanies the day each year when an airplane owner submits his airplane to the maintenance shop for the annual inspection. Some pilots consider it the writing of a blank check, and most of the others realize that something will probably come up during the conduct of the annual that is unexpected, and might put a dent in their wallet.

The manufacturers of airplanes generally provide a checklist in their maintenance manuals for the conduct of annual inspections. For those aircraft that, by the nature of their operations in the commercial environment, are required to undergo virtually the same inspection every 100 hours of flight time, the same basic inspection and checklist is required to be accomplished.

Those of us who have owned an airplane for any appreciable period of time or who have been exposed to the industry for more than a neophyte period realize that a properly performed annual inspection is a very detailed and time-consuming process that increases in complexity geometrically with the complexity of the airplane involved. As brought out in Chapter 3 on aircraft ownership, the expense to maintain a complex airplane is often much greater than that expended for its more simply constructed stablemates.

In order to perform the required inspection and certify in the logs that the aircraft has been satisfactorily inspected and returned to service, the person who is going to make that certification is required not only to possess a mechanic's certificate, but also to have FAA *inspection authorization.* Inspection authorization is not granted to every mechanic, and there are certain requirements that involve a period of experience as a mechanic, as well as more advanced knowledge and testing that must be possessed and accomplished, before a mechanic can seek inspection authorization from the FAA. Because those with inspection authorization (IAs) are deemed to be individuals who possess advanced knowledge and skill beyond that required of the certificated mechanic, the inspection function is one upon which the law imposes some of the highest duties regarding maintenance facilities.

The silent danger of a defect in the airplane that has not yet manifested itself to the pilot is just as potentially lethal as is the problem of which the pilot is already aware: Fuel cells can be leaking internally; high-pressure fuel lines and oil lines can be in an advanced state of deterioration and ready to wreak havoc should they burst or otherwise leak; landing gear systems can have defects or problems waiting to affect their operation. The list could go on for pages. It is up to the IA who is performing the inspection of the airplane to diligently follow the manufacturer's recommendations, as well as the regulatory requirements to inspect the airplane and repair not only those things of which the pilot has made him aware, but also to discover the lurking problems that have not yet developed to the point that the pilot is able to list them on a squawk sheet.

In addition to the necessity that the manufacturer's recommendations for inspections be followed, there is a requirement in the law that the inspector is expected to use *good and workmanlike technique* in the conduct of his inspection.

That requirement of a good and workmanlike manner might mean that the inspector needs to go beyond the bare minimums of the manufacturer's recommendation checklist, and inspect things not listed thereon or referred to in merely a general manner.

The reason the IA has his inspection authorization is that he has shown himself to be sufficiently experienced, and to have advanced knowledge over that required of the mechanic, so that he is a person in whom the trust can be reposed to perform this most vital of function. A great deal of those advanced requirements for receiving inspection authorization will be made known to a court or jury during a lawsuit that results from an accident involving a mechanical problem with the aircraft. It simply cannot be overemphasized that the inspection function is the physical examination of the airplane. Just as the pilot is required to undergo a check of her health on a periodic basis to ascertain that she will likely be able to successfully perform until the next examination, the airplane is required to be similarly examined.

IAs in the field need to realize that the annual inspection is designed not only to uncover and repair what is wrong now, but also to make as sure as possible that the airplane will perform another year. Too many aircraft see a maintenance shop only for the annual inspection, with very little, if any, work done on them between annuals.

If an IA realizes that he is dealing with such an airplane, he needs to be all the more vigilant. If he has good reason to believe that this airplane is not going to be seen again by a certificated mechanic until next year, he should make sure that he conducts his inspection with that in mind. There are probably more components of an aircraft that suffer as much from the ravages of time as there are those that wear out through operation. Hydraulic hoses, fuel hoses, fuel cells, and similar components and systems can become unairworthy over time, even if the airplane is not flown between annual inspections.

In the last analysis, it is the IA who will decide if a particular aircraft is airworthy and can be returned to service, and if it can reasonably be expected to remain airworthy for the next year. Like any other maintenance, if the aircraft owner or pilot wants a skimpy annual, I would recommend that the competent, prudent IA shy away from working on that airplane. It always seems to be that the pilot with that attitude will be involved in an accident sooner or later and then subsequently forget that it was he who tried to put pressure on the IA to pinch every penny that could be pinched during the conduct of the annual inspection. Human nature being what it is, often it is that pilot who will later present a claim against the maintenance shop for the very result that was made likely to happen by his own attitude. The competent and prudent mechanic does not need this exposure—he already has enough problems complying with the duties that the law commands of him when he tries his best to do a thorough and complete job.

Since the passage of the General Aviation Revitalization Act of 1994 (GARA), which is the federal law discussed in detail in Chapter 9, maintenance shops need to be even more vigilant to their potential to be sued when an aircraft accident happens. GARA grants immunity to the manufacturer of most general aviation aircraft after a period of 18 years has passed since the aircraft was first sold. This is the product liability reform law which was championed by almost every segment of the general aviation industry. But, as with everything else in life, it is a compromise.

The increasing costs of product liability insurance, defense counsel, and eventual settlements or court judgments were partially responsible for the disastrous decline of the general aviation industry in the 1980s. Congress finally passed GARA after years of attempts to do so, and intense lobbying for it by manufacturers' representatives, and just as vigorous opposition by plaintiffs' interests and their lawyers.

Since manufacturers are now immune from suit for the vast majority of airplanes in the general aviation fleet, they will no longer be the target defendants of plaintiffs in the lawsuits which follow almost every significant accident. I have been advising clients and industry groups to which I regularly speak to watch and see whether maintenance shops will now have the unfortunate honor to replace manufacturers as the target of avaricious plaintiffs. I predict that they will.

Another, similar situation arises when a shop works on an airplane made by a manufacturer which either is no longer in business at all or has gone through a reorganization under the bankruptcy laws. Even without the benefits and protection of GARA, no plaintiff is going to recover from the manufacturer of a classic Taylorcraft, since that company went under in the early 1950s. Claims against the old Piper are being handled by a trust established pursuant to the bankruptcy proceedings that resulted in the demise of the former company, and the creation of a new corporation.

In these types of situations, the "last wrench" on the airplane might be the only legally viable defendant whom a plaintiff can pursue after an accident occurs. If that is the case, the maintenance facility can expect to be sued, and the plaintiff will come up with some theory of liability to try to pin fault on the shop.

FIXED-BASE OPERATORS

There is no good or generally accepted definition for the often-used phrase *fixed-base operator.* Most general aviation airports have some kind of commercial business established at them, and somewhere in the annals of aviation these endeavors became known as fixed-base operators, or FBOs. The FBO at some airports might be engaged in a wide spectrum of businesses, while those at others might limit their activities to very few areas.

The small municipal or county airport might have just one operator who is in charge of everything. He might be involved in renting aircraft, running a flight training operation, employing mechanics, providing maintenance services to both steady customers and transients, selling fuel, renting tiedown and hangar space to based aircraft, and providing a place of business for public use. Let us look at all these business activities, and the potential negligence liability that accompanies each.

When a fixed-base operator rents airplanes, whether it be to students undergoing training or to customers who are already certificated pilots, the discussion at the beginning of this chapter is certainly applicable to that fixed-base operator. The

FBO who is in the business of regularly and routinely renting airplanes to others has a duty just as high as, if not higher than, the individual who might occasionally rent her aircraft simply to defray the otherwise burdensome costs of ownership of it. The FBO has the same duty to maintain the aircraft in an airworthy condition as does the individual who rents, and the FBO might even be in a better position to ensure the qualification and competence of renter pilots.

While it might be difficult for the individual who rents her airplane to be vigilant to discover the true level of qualification and competency of her rental customers, the FBO would seldom have an adequate excuse. If the FBO is in the business of renting aircraft, it is almost certain that he is going to have flight instructors on staff, or with whom he is closely associated, who should certainly be able to perform checkrides and, as elsewhere discussed, do what is necessary to delve into the qualifications and competence of those who come to the operation to fly rental aircraft.

FBOs can be found liable for the negligence of a renter pilot in a couple of ways. Some states have laws on the books, or court decisions as a part of their common law that imputes the negligence of the pilot to the one who rented the aircraft to the wayward aviator. If you are in such a state, and rent an airplane from an FBO, that FBO can be held responsible for accidents that you may suffer, if caused by your negligence or, naturally, the negligence of the FBO itself.

Most all the states recognize the principle of negligent entrustment. This legal theory says that it is negligent on one person's part to entrust certain instrumentalities to another who is not capable of using them properly. That long-winded explanation is fairly easy to illustrate. Give a firearm to a 5-year old, and you'll be responsible if the child shoots someone or damages property with the gun.

The same result often occurs if a car dealer turns a 16-year old loose in a fast car for a test drive, and an accident ensues. Each state which recognizes this separate wrong, negligent entrustment, may have some differences in the elements that a plaintiff must prove to hold the owner of the entrusted property liable for the negligence of the renter or borrower. Basically, all the states which have negligent entrustment as a part of their bodies of law require that the instrumentality involved be one which has a high capacity for danger in the hands of the incompetent, and either knowledge that the user is incompetent to use the property safely, or a lack of any reasonable attempt to find out the skill level of the borrower or renter.

If an FBO rents aircraft to any warm body who can produce a valid pilot and medical certificate, that FBO is playing close to the edge of negligent entrustment. Checkrides serve several purposes when a new renter pilot comes onto the scene at any FBO. One of them ought to be to determine this person's skill levels. Once that is known, the FBO should act accordingly in renting only aircraft that the particular pilot is able to fly well, and perhaps imposing other restrictions upon the renter's use of the FBO's airplanes. Many a good FBO will set standards for such things as minimum weather (above the levels set forth in the FARs) for both local and cross-country flights. These FBO's standards should be in writing, and should be a part of the rental agreement that the pilot is asked to sign.

Every rental operation should consider a written rental agreement as a required piece of paperwork. A well-drafted agreement doesn't need to be overly long or verbose, and it also doesn't need to be composed in legalese. It must set forth the obvious—rental rates and payment requirements. Well-done ones go more deeply into the FBO's rental policies and requirements. Some FBOs restrict rental pilots from flying into sod fields, but I've always doubted the wisdom of such carte blanche prohibitions. There are many fine sod fields, and using them should be a part of every experienced pilot's education. However, the intended field should be approved by the FBO, or by one of the instructors.

Weather minimums should be different for day and night flights. No pilot can safely fly cross-country at night in the same weather as can be safely tackled in daylight. A good agreement will also set forth maximum crosswind components for rental flights in the FBO's airplanes, even if the operation is located at an airport with multiple runways, where crosswinds usually aren't a major problem. Destination airports, where the renter is going, may well be single-runway fields. Of course, at some point, a maximum wind velocity should be established, regardless of direction.

Doing all these things, enforcing the standards, and requiring thorough initial and recurrent checkrides will go a long way toward establishing that the FBO does not negligently entrust its airplanes to others.

There was a time, decades ago, when many courts held airplanes to be inherently dangerous instrumentalities. Thankfully, that is no longer the case in any state of which I am aware. But judges who are not themselves oriented toward general aviation usually think of "little airplanes" as risky contraptions, and sometimes doubt the basic sanity of those who fly them. Guard against this silly notion, and have the documents in place, properly signed, that show that the FBO uses due care in renting aircraft.

Many states also require a written notice to be given to a renter pilot which discloses the terms of the insurance coverage which the FBO has in place for rental operations. Every rental agreement should contain this information, regardless of whether the state laws actually require it. Many a renter pilot will have an accident, and then complain that she was unaware that her own liability wasn't covered, or that she didn't know that the FBO's insurer had a right to go against the renter to recover the money that the insurance company laid out to fix the airplane, or pay other damages.

Renter pilots should read the agreement carefully, and understand it before signing. If you fit into the majority of folks who aren't very conversant with insurance terms and policy language, find out the details before renting. Call your own agent or broker, speak with the FBO's insurance representative, or seek the advice of an attorney. Just don't go flying, having no idea in the world of the risks for which you're covered, or not covered. Worse yet, don't assume that you have coverage that is just not there.

During the last several years, but particularly in the last decade, the *sale-leaseback* has become a very popular means for FBOs to place aircraft into their rental fleets. The transaction is not a complicated one. The FBO, or someone else, sells an aircraft to an individual who then leases that aircraft back to the FBO for rental use. While this is obviously not a book about tax law, in the days before the Tax Reform Act of 1986 (TRA 1986), it was extremely advantageous for higher-income individuals to procure as many tax shelters as possible, to shield their other income from dissipation by federal income taxes. TRA 1986 did reduce the usefulness of many tax shelter devices that were previously quite popular, and included among those is the aircraft leaseback transaction. But there are still, without the formerly obtainable tax breaks, some good reasons that this business can and should continue.

It goes without saying that airplanes are extremely expensive to buy, particularly new ones. Very few individuals who do not have a specific need for frequent business transportation by general aviation aircraft are in a position to buy a new or late-model airplane that is well equipped with avionics and other options. Most persons who would like to own such an airplane need some sort of assistance in terms of an income stream to be able to make the acquisition and operation of the aircraft an affordable experience. The leaseback does provide that.

When an aircraft owner leases his aircraft back to an FBO for the FBO to then use in the rental fleet, or for some other commercial purpose, the owner who is engaging in the leaseback must be aware that he is taking on new exposure to liability that he would not have but for the leaseback arrangement.

If the aircraft is involved in an accident that can at all be related to its condition, maintenance, or lack thereof, the owner can expect to be sued right along with the pilot, FBO, and virtually anyone else who is within reach of some theory of liability. The leaseback owner needs to be absolutely certain that he is listed as a named insured on the FBO's insurance policies, and that the coverages and limits offered under those policies are sufficient to protect the FBO and also to insure the airplane owner against the exposure that he might reasonably face.

Before going into a leaseback arrangement, the prospective owner-lessor needs to take a good, hard look at the realistic dollar numbers of the proposal. I know of few situations in which leaseback income has actually paid the total costs of owning and operating the aircraft. The most that can realistically be hoped for, in almost every situation, is that the cost of ownership will be somewhat reduced, and that over a period of time, the capital acquisition cost of the aircraft will be recouped and paid for by the income from the leaseback. Seldom, if ever, is the airplane flown enough to cover all the costs of acquisition, financing, insurance, maintenance, and overhaul reserves that are incurred during the period of time the FBO has it in his rental fleet.

A leaseback owner should realize that rental pilots are not going to treat his pride and joy with the same kid gloves that he would use in his own flying, because a rental airplane is a rental airplane. The renter pilot does not really know, or care, whether the aircraft actually belongs to the FBO, some institutional investor, or an individual who is looking to reduce his own costs. While rental airplanes are not particularly abused in the industry, they are—often through the inexperience of student pilots and newly certificated pilots who fly them—subjected to abuses that an experienced, well-intentioned owner would not permit. For some reason, rental airplanes always go through brakes faster, wear out more tires, seem to be in the radio shop more frequently, and suffer other, higher maintenance costs than an airplane that is used by an experienced pilot in the course of her own flying. This is to be expected, though, and the rental price that the leaseback owner receives from the fixed-base operator must be sufficient to cover these costs. If it is not, the leaseback owner needs to know that and make his plans accordingly.

The leaseback is often attractive because, with the right situation and the right fixed-base operator as the lessee, the aircraft owner can see his airplane fly enough to, while not being profitable, at least provide enough cash so that the owner can afford to own an aircraft that he otherwise could not.

While on that subject, no leaseback is perfect, and like many other facets of the aviation industry, leaseback arrangements tend to be temporary. Unfortunately, fixed-base operators go out of business or, as a rental aircraft gets older, its attractiveness to renter pilots declines, and the FBO will use its option, which is contained in most leases, to cancel upon relatively short notice, and cut the owner adrift. At that point, the owner is left with an older airplane, sometimes with quite a few more hours on its engine and airframe toward an upcoming overhaul, and with no leaseback income to pay for the continued ownership of the airplane or the refurbishment or overhaul which it might soon need.

I never recommend that someone buy an airplane that he cannot otherwise afford, depending upon leaseback income to pay for it. A leaseback should be considered a way to reduce the cost of ownership of an airplane, but should never be depended upon to carry its own weight.

Should you be approached by an aircraft broker or fixed-base operator to purchase an airplane and enter into a leaseback arrangement, consider it studiously and pay attention to the dollars and cents of the transaction. If you are not sufficiently experienced, ask the FBO for the names of several other leaseback owners in his rental fleet so that you can contact them. Does the FBO pay the hourly payments to the owners on time? Does he maintain the airplanes well? Does he overmaintain the airplanes, using his maintenance shop as a source of profit for himself at the expense of the leaseback owners who are paying the maintenance bills? There are numerous other questions that need to be asked, and the best audience for them is undoubtedly another leaseback owner or two with whom this FBO has done business.

Then, the purchase of the aircraft and the lease itself need to be handled professionally. Refer to Chapter 3 concerning many of the points involved in the purchase of an aircraft. The *lease is a binding and obligatory contract* between the owner, who will be the *lessor,* and the FBO, who will be the *lessee.* That document will create *rights and obligations* for both parties, just as any other contract does, and needs to be negotiated with the input and advice of someone who is familiar with the industry. The aircraft owner should not merely take the form of lease that is put in front of him by the FBO and sign it. There are many provisions within it that need careful examination by an attorney experienced in aircraft leasebacks.

Many questions—such as those surrounding the insurance coverages required, the cancellation provisions, the maintenance requirements and options by the FBO, and requirements concerning where the aircraft is to be maintained and where fuel is to be purchased—need to be carefully negotiated and thoroughly considered prior to execution of the leaseback.

While most FBOs are ethical and upstanding operators, there are, unfortunately—just as in any other industry—those few who structure their transactions solely for their own benefit and at the expense of those with whom they do business. No leaseback owner wants to be on the other side of those types of transactions, but without competent advice and careful deliberation, an inexperienced aircraft owner might end up in such a position.

As previously stated, leaseback owners need to be aware that they will probably be sued, should their aircraft be involved in an accident wherein any serious injury or property damage occurs. Again, the primary answer to this dilemma is to purchase the correct insurance in the proper amounts to give some peace of mind to the owner of the airplane. But just as important as insurance is the competency of the FBO to whom the airplane is to be leased. If the FBO runs a first-class operation and rents airplanes only to those pilots who are qualified and competent, and as well operates a flight school in the manner in which it should be run, a great deal of the distance has already been traveled toward a safe and secure operation.

Another activity in which fixed-base operators suffer exposure to negligence liability is the potential to be called into court to answer for the negligence of flight instructors they employ. Centuries ago, the courts were presented with a common problem, and fashioned a unique answer. Somewhere back in the Middle Ages, one serf driving an ox cart undoubtedly ran into another. Because in feudal Europe serfs were only slightly removed from the status of slave, the serf who was at fault was hardly able to respond monetarily for the damage he did to someone else's ox, cart, and cargo. So the judges of old England decided that the best way to solve this problem was to fashion a system whereby the employer, then called the master, would be liable for the negligence of the employee, then called the servant.

So today we have the principle that when an employee commits a tort, in this case negligence, while within the scope of her employment and in the furtherance of her employer's business, that employer is liable for that negligence committed by the employee. The old-fashioned term *master-servant rule* is still used in many legal works, and by some lawyers and judges, to describe this arbitrary, but workable, legal theory that is very much with us today.

The term used in more modern and artfully drafted opinions of courts is *vicarious liability.* This phrase has a definition that is broader than necessary for the purposes here, but vicarious liability is generally a principle whereby one person is monetarily liable for the actual wrongful conduct of another. Hence, the master-servant rule comes within the scope of vicarious liability.

To boil this all down into common sense, we must simply remember that in today's legal world the employer is going to be fiscally responsible for the negligence of the employee. The basic exceptions to this are only twofold. If the employee commits the tort (negligence) while acting outside the scope of her employment, the employer has a good defense. The other shield the employer may throw up in an attempt to avoid liability is to argue that the person was not an employee at all, but rather an independent contractor.

Independent contractors are liable for their own torts, as are employees, but do not thrust that liability upon the person who has engaged them, as an employee does upon his employer. The line of demarcation between employee and independent contractor status is not clear, and is often blurred by lawyers and courts. The most important of the basic differences becomes one of how much control the employer has or may exert over the performance of the assigned task. Another of the controlling factors is which person, the employer or employee, furnished the tools with which the work is to be done. In the arena of flight training, the FBO provides the airplane, sells the ground school materials, and furnishes the place of business. Therefore, control, not who furnishes the tools, becomes the point of debate in most of these cases.

If a person commissions a portrait artist to paint a portrait of that person or a member of his family, very few patrons would presuppose to tell the artist how to do his work. This is an example of an independent contractor arrangement. Conversely, when an employed gardener reports for work, begins the tasks using the employer's mower, trimmer, shears, and other equipment, and is told what bushes to trim, how to trim them, how short to trim them, in what order to trim them, and when to be done with the work, that control on the part of the employer places the gardener quite easily within employee status.

What does all this mean for an FBO? Simply that the employer, who is the FBO, whether it be an individual, a partnership, or a corporation, is vicariously responsible for the torts, including negligence, of its employees. These employees can be flight instructors, mechanics, line service operators, snow removal personnel—the list is endless.

If a Certified Flight Instructor (CFI) is negligent in the conduct of giving flight instruction, and plenty about that has already been said in this chapter, that CFI will certainly bear liability toward the injured person or persons. In addition, vicariously, his employer, the fixed-base operator, will be hauled into court with him. Vicarious liability does not replace liable persons, it simply adds the employer in addition to the employee whose wrongful act caused the harm in the beginning. In other words, the flight instructor and the FBO are both liable, and the plaintiff can get a judgment against both, and collect that judgment from whichever has the assets to satisfy it, or, more appropriately, whichever has the insurance to satisfy it.

Much has already been said about the liability of fuel suppliers and maintenance operators. Generally, those functions are performed by fixed-base operators and, therefore, FBOs who do that work are liable for the consequences of negligent or other wrongful action or inaction. As with the situation with the flight instructor, the negligent mechanic or IA will not face the music alone, but the plaintiff will bring the FBO into court as well. Again, bear in mind that the best solution to this paradox is the purchase and maintenance of adequate insurance with broad forms of coverage that are needed to cover the activities in which the FBO engages, and which has limits sufficiently large to cover the liability to which that FBO is likely to be exposed.

Almost every FBO qualifies, as well, for another invitation to come into court. Most FBOs maintain some kind of control over certain parts, if not all, of the premises of the airport at which they do business. They are therefore in a classification that is known in the law as *keepers of premises.* What this means is that individuals or businesses that own or occupy (as a tenant) real estate, which is called the *premises,* are potentially exposed to liability to persons who suffer injury or other harm on those premises as the result of some legally recognized wrong committed by the keeper of those premises.

Most states look at an individual who comes onto the premises of another as being in one of three classes: trespassers, licensees, or business invitees. In the area of real estate law, into which we are now delving, there is a significant difference in the law among the various states. It is especially important to seek competent legal counsel in your own jurisdiction, and never rely totally upon general principles without the opportunity to find out how those general principles have been modified or abrogated by the courts where the reader lives.

Generally, the occupiers of land owe no duty to a *trespasser* other than to refrain from willfully causing him bodily harm if the circumstances do not justify bodily harm. Again, the various states of our country have extremely divergent views regarding if, when, and to what extent bodily harm can be rightfully used against a trespasser. Get good advice before you ever think that you are justified, in your state, in using any force against a trespasser. But the point, as far as this work is concerned, is that in most states, the occupier of premises does not have any duty to take any steps to protect the safety of a trespasser. One modification of this rule, in effect in most states, is the *attractive nuisance doctrine.*

Courts in the states that have adopted the attractive nuisance doctrine have justified it on the basis that small children of tender years are incapable, by virtue of their very young age and immaturity, to recognize, appreciate, and avoid dangers that are attractive to them. Therefore, in these states, young children who are, nevertheless, trespassers, are treated differently from more mature trespassers.

The attractive nuisance doctrine holds the occupier or keeper of the premises liable if a child of tender years comes to grief through an attractive nuisance such as a swimming pool, a dog kept upon the premises, machinery left with the key in it (such as a bulldozer or airplane), and similar situations where, when an objective view is taken, it is easy to see that a young child could easily be enticed to danger through no actual fault of her own. So the FBO must be aware of attractive nuisances about his airport—and there are plenty of them; they must be adequately secured or removed.

Once we leave the trespasser classification, the next class into which some individuals fall who come onto the land of another is that of a *licensee.*A licensee is a person who has permission to come onto the land, but is not coming on the land to further any commercial or business purpose of the occupier of the land. The invitee, therefore, is a social guest. In most states, the occupier of the premises has the duty to not willfully injure the invitee, and as well, to make the invitee aware of any particular dangers of which the occupier is aware.

But the keeper of the premises is not generally responsible to be diligent in any attempts to ascertain dangers in this situation. So toward the social guest, the keeper of the premises owes a duty not to injure him and to simply make him aware of whatever dangers there are of which the keeper of the premises is aware. As far as fixed-base operators are concerned, there will probably not be any persons who fall into this category who come onto their land. This is because a fixed-base operation is a commercial activity and the courts will hold that people who come onto its premises with permission will be coming as either actual or potential customers, and therefore are a class of individuals that is coming onto the land in furtherance of the FBO's business.

These people who come for commercial purposes are known as *business invitees,* or merely *invitees.* The occupier of land owes the highest duty of all three classifications to his business invitees. He not only owes the duty that is owed to a licensee, which is to not willfully injure him and to make him aware of known dangers; but as to the business invitee, the business owes the duty to reasonably discover all dangers and either eliminate them or, if they cannot be eliminated within a reasonable period of time, issue an effective warning that communicates the presence and nature of the danger to that business invitee. Now it should be understandable why, when someone slips on the broken egg at the grocery store, she can recover damages for whatever injuries ensue in that accident.

An FBO's premises usually contain two basic areas of activity. First is what I will call the *nonflying area,* or *ground area,* and the second is the *flying area.* The ground area includes the auto parking lot, the terminal or office building, any walkways between them, and all other areas over which the FBO has control that are not open to aircraft movement.

In these ground areas, the FBO has the same types of duties as do other commercial businesses not associated with aviation. The parking lot needs to be kept safe, snowfall needs to be removed from the lot and exterior walkways, holes in the sidewalk need to be repaired, and all the other maintenance of buildings and grounds, both inside and out, needs to be done by an FBO to the same extent as others engaged in business. Unfortunately, the FBO's exposure does not end here.

In the *flying area* of the airport premises, the exposure to liability is far greater than in the ground area. It seems as though airplanes come to grief all too frequently because of the negligence of fixed-base operators in maintaining their airports.

Any of us who have flown or have been otherwise involved in this industry for any appreciable period of time know of situations where landing gear have failed because of holes in runways, or aircraft have hit piles of snow that were too close to runways and taxiways; and most unfortunately, it is still not that rare for individuals to be severely injured, if not killed, by walking into moving aircraft or propellers.

A little dose of good sense goes a long way toward solving the liability problem for FBOs in the flying area of the airport. The first rule is that people who do not belong there must be kept out. Airplane watching is a fun endeavor for most people who do not themselves fly. Certainly this is one avenue to attract people to the airport. Some of them might become sufficiently interested to go into the office and start asking questions, and perhaps become customers for the FBO. I definitely do not mean to discourage that; but I propose that airplane watching must be controlled. Just as good fences make good neighbors, good fences do not need to hinder the pleasure of the public in watching the goings-on at the airport and can, at the same time, provide the necessary level of protection for both the observer and the FBO.

Next, the flying area must be kept in decent condition. Tiedown ropes need periodic inspection and replacement. All too often an airplane will be damaged in a windstorm simply because the tiedown rope broke because it was old, rotten, or otherwise ready for the trash pile. The runway needs to be maintained. Perhaps the locals know where that chuckhole is, but how is the transient pilot to discover the hole before it is too late? If grass areas of the airport are unsafe for taxi, mark them with a sign that says so.

Think ahead, and try to anticipate what can happen if everything goes to pot quickly—because someday it invariably will. The FBO who thinks ahead, plans for the unlikely, and tries to foresee areas where people can get into trouble, both in the ground and flying areas of the operation, will probably prevent that trouble, and be far ahead of the individual who does not plan but who lets events control him.

Lastly, in the area of premises liability, the FBO needs to make his insurance agent or broker well aware of the extent of the premises over which he has control. Then the insurance agent or broker can advise the operator on the forms and limits of coverage that are needed to protect him from reasonably foreseeable levels of liability exposure.

PILOTS

It is an unalterable fact that whenever an aircraft accident occurs, a pilot is involved. If someone is injured, killed, or suffers a loss to his property in today's litigious society, he is very likely to be looking for someone upon whom he may lay the blame for the accident, in hopes of gaining some recovery. In almost all circumstances, the pilot of the airplane will at least fall into the classification of those who will be seen to bear some, if not all, of the liability for the accident.

Pilots can be negligent in a myriad of ways. It would take hundreds of pages to try to explore them all; and then, without doubt, one or more would be omitted. Most of the aviation journals have some coverage of recent accidents wherein one or more authors will attempt to carefully examine, dissect, and then explain a particular aircraft accident. For as many years as I have been reading these articles, I have yet to see exactly the same accident discussed twice, and have always learned something of value from each such reading.

Pilot negligence is a difficult thing to quantify. Often the error that occurs and causes the accident is actually the culmination of several errors or events that, if viewed separately and independently, were small situations, but when acting in concert, overpowered the skill of the pilot. Many accidents begin with the preflight planning, or lack thereof. It cannot be too greatly stressed that one way to avoid an accident is to avoid the small mistakes that seem to mushroom into larger problems.

After an accident has occurred, investigators, lawyers, and courts will have years to contemplate a situation with which a pilot might have had to concurrently contemplate and deal with in a matter of a few seconds. Also, hindsight is always 20/20, and foresight does not benefit from that unfailing acuity.

The presence of liability exposure should not frighten any competent pilot from the practice of his vocation or avocation. Mistakes are unavoidable, that is why one of my personal clichés is that a piece of rubber is on the end of every pencil for good reason. But when a mistake results in an aircraft accident, it is almost certain that a lawsuit will follow.

Pilots are referred to the discussion in Chapter 5 concerning aviation insurance. Adequate insurance is absolutely necessary—in my humble opinion—before a pilot even starts the engine of an aircraft, let alone flies it. To fly without that insurance puts everything the pilot is and has at risk. Only a fool makes light of the fact that all humans make errors, and someday an error can bankrupt even the most successful of ordinary individuals if they go about their activities without proper insurance. Also, it is an unfortunate reality that aviation accidents tend to have "larger numbers" attached to them than do automobile cases; airplanes are capable of doing more damage. The people who ride in general aviation airplanes are often of a higher income level than the average citizen; and therefore, when they are severely injured or killed, the damages suffered by them or their estates are higher than in a more typical lawsuit.

It is this fact that accounts, to some degree, for the level of judgments and settlements in aviation cases. There are many, many reasons that the value of injuries and the damages attendant to wrongful deaths have increased so greatly over the years. But in our industry, most people who have the financial wherewithal to fly, own, or rent airplanes—and the people with whom they tend to associate, their passengers—are members of the higher socioeconomic sectors of society. So when the average exposure in an aviation case is scrutinized, it is not a like comparison to the average exposure in an auto accident suit. Sure, upper-income people are injured in auto accidents, but it is just a fact of life that their numbers tend to be higher in aviation cases because those individuals of higher economic means are customarily a part of the general aviation population.

Therefore, pilots need to take a good, hard look at the limits of coverage that are provided by their aviation insurance policies. Again, reference to Chapter 5 is in order, but more importantly, a frank discussion with an aviation insurance agent or broker and an attorney who is knowledgeable in these areas can give you guidance in selecting the levels of coverage that are necessary to serve your needs and to protect what you have from the ravages of an uninsured or underinsured accident.

FLYING CLUBS AND OTHER CO-OWNERSHIP ARRANGEMENTS

Chapter 3 discussed the forms of co-ownership of aircraft; there was a limited discussion of the liability of co-owners in each of the organizational forms covered. This final portion of Chapter 8 explores that material more deeply, and offers suggestions to minimize liability exposure.

First, let us deal with the informal *co-ownership arrangement* among individuals. As a matter of brief review, this is a transaction in which two or more people purchase an airplane as joint purchasers, register the aircraft as joint owners with the FAA Aircraft Registry, and have their joint names appear on the certificate of registration. If the aircraft is involved in an accident that can be attributed to the fault of the pilot, it is likely that his co-owners will be named as defendants in a lawsuit if one arises from the injuries or other damages caused by that accident.

One principle of the law that can drag co-owners into lawsuits for which they have no practical fault is a *doctrine of joint enterprise* or a *common enterprise.* If the courts find that there is a common business enterprise going on by and among co-owners, those otherwise innocent co-owners can find themselves liable for the negligence of the other owners, whose fault caused the accident. Likewise, if the aircraft is itself defective, improperly maintained, or for some other reason unairworthy, it is likely that a plaintiff will try to establish that all members of the common enterprise had a duty to ensure the safety and airworthiness of the aircraft, and that that duty was breached when the accident occurred.

The informal co-ownership arrangement between individuals is, of the various forms of co-ownership of an aircraft, the simplest, and easiest to form. In combination with the discussion in Chapter 3, though, it should now be seen that informal co-ownership is not the preferable way for groups of individuals to enjoy the benefits of aircraft ownership; it is simply fraught with too many problems. Seldom is there an arrangement worked out to solve the problem that invariably will arise when one of the co-owners wants or needs to divest himself of his interest; more importantly, the co-owners are quite possibly exposing their entire personal fortunes to liability for the wrongful acts of another. Very few of us want to put ourselves in that position, and once we intelligently understand the risk involved, we tend to shirk from it.

So long as enough insurance is carried, I suppose that the issue of personal exposure for someone else's acts could be seen to be only an academic argument. But who is to decide when insurance limits are so high that the losses covered in an accident could not conceivably exceed them? Certainly not I.

If a flying club is organized as an *unincorporated association,* it still runs the risk that the veil of protection provided in most states by that status can be pierced, and with varying ease from jurisdiction to jurisdiction. Unincorporated associations do not give protection from personal liability that a true corporation renders to its shareholders. Therefore, if your flying club is an unincorporated association and you have an ownership position in it or in the airplanes used by it, it is not unforeseeable that in the event of an accident the owners of the club and of the airplane(s) could find themselves as personal defendants. If your flying club is organized as an unincorporated association and you are so involved in it, be absolutely certain that the club's policy of insurance extends coverage to the individuals involved. You need to have the policy reviewed by a competent insurance agent or broker and an attorney familiar with the forms of aviation insurance policies.

The danger is the same as always. The person who causes the accident, whether that is the pilot, the mechanic, or someone totally distant from the flying club, cannot control the course of the litigation. Neither will that person take the total fall by himself if his attorney is able to find other pockets with which to share the misery. All flying club owners and participants need to be aware that an unincorporated association is not a very strong shield against personal liability and must deal with their club, their fellow members, and the third-party outside world accordingly.

If the flying club is organized as a *true corporation,* and if the corporation is properly maintained, a great deal more protection is gained for the individuals involved in the club. Proper corporate maintenance means that the required form of doing business must occur. Shareholder meetings must be held, the board of directors is required to meet, officers and directors must perform their respective functions properly, and good records need to be kept of all the corporation's activities. A corporation is absolutely no good at all if it is not properly maintained, or if the individuals involved in it do not recognize it as a separate legal person, but see it as an alter ego and deal with it as such.

This is certainly not a complete guide to corporation law, and when a flying club is incorporated, it needs the services of a competent attorney to advise it, not only in setting up the entity, but also in working with the members, directors, and officers to function and act like the true corporation that the club is. In this way, when the need arises for the protection from personal liability, that protection is there.

A corporation is seldom formed for someone else's benefit. The same goes for a flying club—the reason they are incorporated is usually to provide the limited liability for the shareholders that is gained through corporate existence. But this limited liability will evaporate if the corporation is not properly operated and the individuals involved do not deal with it as the separate legal person that it is.

If the corporation is properly organized and maintained, it might well be liable for the fault of someone else who has an accident. But if the corporation's assets are kept at a minimal level, and if it does other things correctly in its financial planning and business, it can mount a formidable defense to an attack, particularly one that seeks to pierce that corporate veil and go after the individual shareholders' personal assets. Thus, the shareholders might see the assets of the corporation at risk, but their own individual holdings should be protected if they had no direct involvement with any factor causative of the accident.

9

Product Liability

THE LIABILITY THAT MANUFACTURERS FACE HAS BEEN CITED AS ONE of the primary causes for the decline of the general aviation industry into an almost depressive state during the early to mid-1980s.

Peak production of general aviation aircraft, when determined by measuring units manufactured, occurred in 1978. Since then, the industry has endured a precipitous slide, with aircraft powered by single piston engines seeming, for a while, to be headed for the same fate as other extinct birds. As of this writing, Cessna Aircraft Company has not made a piston engine airplane since 1986, but the company is now gearing up to make some of its former single-engine models again, including building an entirely new factory in which to do it. Cessna refused to reenter the single-engine business until the law of product liability changed significantly. When you get to the discussion of the General Aviation Revitalization Act of 1994 (GARA) later in this chapter, you'll see why Cessna was motivated to restart production of single-engine airplanes. The first of the new models is scheduled to be built in 1997.

Piper Aircraft Corporation went through a change of ownership in the late 1980s, from being part of a large conglomerate to being an individually owned company in the hands of a person who directs it as the sole shareholder, director, and chief-cook-and-bottle-washer. As business worsened, and the owner chose to forgo product liability insurance, bankruptcy was inevitable. Piper operated under the protection of Chapter 11 of the bankruptcy laws for several years. In 1994, it emerged from bankruptcy as a new corporation, called The New Piper Aircraft, Inc.

Beech Aircraft Corporation continues as a part of another large conglomerate, but its production of piston engine singles was limited to only two models in 1996, and not many of those. While still manufacturing one piston twin, few units are sold each year, and Beech's primary efforts seem to be more and more directed toward the turbine market, as Cessna's were exclusively before it moved back into piston engine production.

Mooney Aircraft Company is a bright spot in the industry, along with Piper. Mooney continues to introduce new models, which in essence are continued refinements of an already well-proven product line, and as of this time, concentrates its efforts solely on the single-engine market.

There is a smattering of other manufacturers, and rumors continue to circulate of resurrection of one or more types of aircraft that are out of production or entire product lines from manufacturers who have either gotten out of the business or gone out of business. Perhaps the worst of the decline of the general aviation industry is past—but only time will tell. There is no question that the number of units being manufactured in the mid- to late 1990s is only a slim shadow of the number that rolled out of the production lines at Wichita, Vero Beach, Kerrville, and elsewhere two decades earlier.

Only a person perfectly clairvoyant and all-knowledgeable could discern all the causes for this sorry state of affairs. But without doubt, the product-liability issue has had a bearing; and depending upon to whom you listen, and in whom you place belief, the degree of responsibility to be put at the base of the product-liability issue varies widely. Plaintiffs' groups and other consumer activist organizations maintain that product-liability suits and claims have not increased that greatly in the last two decades and that the groups and organizations are addressing defects in design and manufacture that should have long ago been eliminated by competent and caring manufacturers.

Defense groups, conversely, hypothesize that the large number of suits and claims is crippling the cost of producing the small general aviation aircraft. A few years ago some costs were released by one of the manufacturers that put the price of product-liability insurance at slightly less than one half of the selling price of a new single-engine aircraft in its model line. Whether that is a true state of affairs is a query to which very few have an accurate answer.

But one thing is for certain. With a large fleet of general aviation airplanes flying, and in a time of declining production, each new unit being manufactured was seen by the bean counters to be required to bear an ever-increasing cost to provide the product-liability protection for the fleet that is already in the field.

Before the enactment of GARA, if there were 50,000 units manufactured by one particular company still active, and that manufacturer is able to build only 1000 units during a certain year, each of those 1000 units is burdened, by some accounting practices, with the costs of providing product-liability protection for the manufacturer's complete exposure, which includes not only the 1000 units to be built during that year, but the 50,000 that are already in the fleet. While I am no accountant, in one sense this appears to be an appropriate method of cost accounting. But in another sense, it also drastically increases the cost, and therefore the selling price, of those airplanes being manufactured and presented for sale during a period of time when they are hard to sell in the first place.

Again, no one has all the answers, and the intelligent and sophisticated can analyze the situation for themselves. It is not the intent of this work to hypothesize the cause of the general aviation depression of the 1980s, but merely to relate what the law is, and to allow each reader to formulate for himself his opinion on the impact of that law on other, greater things. With that in mind, let us begin a discussion of this most important and crucial area of the law, particularly as it seems to be affecting the aviation industry.

A product-liability claim is a lawsuit that seeks to recover damages from a manufacturer or supplier of a product for the death, physical injury, or emotional distress to a person or physical damage to property other than the product in question.

Generally, these cases allegedly arise from the design, formulation, production, construction, creation, assembly, rebuilding, testing, or marketing of a product; and any warning or instructions, or lack thereof, associated with that product; and the failure of a product to conform to any representation or warranty given by the manufacturer or supplier about it.

While general aviation aircraft are obviously products, the definition of a product that is subject to a product-liability claim is far more all-encompassing. Basically, *a product is any object, substance, mixture, or raw material that constitutes tangible personal property, and is capable of either delivery itself or being assembled whole in a mixed or combined state, or as a component or ingredient.* A product is also *that which is produced, manufactured, or supplied for production into trade or commerce, and is intended for sale or lease to persons for commercial or personal use.*

Some additional definitions might be helpful. The product that is the subject of a product-liability case is required to be produced, manufactured, or supplied for production into trade or commerce. It is not something that a manufacturer or supplier builds intended for its own use. The very essence of the public policy behind product-liability law is the thought that *those who manufacture and place products into the stream of commerce ought to bear the burden of defective units, and compensate those who are injured by those units.* This philosophy obviously does not apply, then, to products that are not placed in the stream of commerce or, in other words, to products that are not sold to the consuming public.

Therefore, an aviation manufacturer might be liable for an airplane that it manufactures and sells to the consuming public; but a product-liability claim could not arise from a situation involving a person injured in an experimental aircraft or in a nonexperimental aircraft still in the preproduction stages.

NEGLIGENT DESIGN AND MANUFACTURE

A manufacturer has a duty to use due care in the design and manufacture of its product. The standard of care requires the manufacturer to use that degree of engineering and other design skills as are generally accepted in the industry at the time of the design of the product. Likewise, manufacturing operations must be conducted in compliance with a standard of care, and that standard is, in most instances, a very stringent requirement to manufacture the product according to the design specifications.

When it comes to general aviation aircraft, most product-liability cases involve an allegation that the aircraft was negligently designed or manufactured, or both. The process of designing an aircraft from the drawing board through to a production model is far beyond the scope of this work, and frankly, far beyond my own practical knowledge. Nonetheless, most of us realize that the design of virtually any aircraft involves many trade-offs regarding weight, speed, appearance, and other factors, none of which can be culminated in the design to the ideal extent, and all of which must be balanced against each other in the final product.

Most of the negligent design claims revolve around the flying or handling qualities of aircraft. While not a small subject, handling and flying qualities are, to a great degree, regulated by the applicable provisions of the FARs governing the certification of various types of aircraft. But, as we all have seen, airplanes can become certified, and comply with the regulations, while at the same time having handling and flying qualities that might not be quite as pleasant, predictable, or safe as most pilots expect modern aircraft to be.

The manufacturer certainly has a duty under negligence law to design an aircraft to be safely flown by the average pilot. One should not have to have the ability of Chuck Yeager or the skill of a modern test pilot to be able to safely fly a general aviation airplane from point A to point B. Rather, this airplane needs to be so designed that the average pilot, who possesses only average skills, can safely keep up with it, considering the performance category into which the particular design fits. This is not to say that everyone should be able to fly every design.

Certainly there are many pilots in the population who can very safely navigate about the sky in a Cessna 172 or a Piper Cherokee but who simply do not have the necessary ability to ever upgrade into certain business jet equipment. Rather, the particular airplane, when viewed as a member of a performance class of aircraft, should have design, handling, and flying qualities such that the average pilot who flies similar aircraft can fly the one under consideration without undue safety compromises involved.

If the aircraft is designed for IFR flight, which almost all are today, an entirely different set of considerations also must come into play. Will it be eventually certified for flight into known icing conditions? If so, the flying and handling qualities have to be thoroughly explored when the aircraft is in such conditions, and when a load of ice is carried. In the past, aircraft have been designed that exhibit totally predictable and placid handling and flying qualities until they become burdened with ice. With the attendant differences in the shape of the air foils, center of lift, and weight associated with icing flight, these formerly docile aircraft become virtual tigers to be tamed.

One of the unfortunate parts of this entire discussion is the plain fact that no manufacturer can adequately foresee all possible conditions into which the airplane might be put by those who are flying it. A part of the manufacturer's duty is to foresee, and design against, reasonable misuse of the product, as well as proper use. The philosophy of this quirk in the law is that not only will consumers use the product, but it is human nature that consumers will misuse it. A classic example in this area is the difference between a modern power lawn mower that is purchased today and one manufactured 20 or 30 years ago.

In the last 10 years or so, a requirement has been imposed by the Federal Trade Commission that power lawn mowers be equipped with a safety handle that operates such that, when the latter of two hands is removed from the normal handle of the lawn mower, the engine stops. This is obviously a design feature intended to prohibit those shorter than Kareem-Abdul Jabbar from reaching down to the discharge chute and removing a clump of wet grass while the engine is still running. The manufacturers of power lawn mowers are well aware of how many fingers and toes are lost each year in the blades of their products. Hence, they have designed around that foreseeable misuse, as the law requires. The same requirement exists for airplanes.

Obviously, manufacturers cannot use the same degree of latitude in designing against foreseeable misuse when considering an airplane design as they can when formulating lawn mowers. So far, the courts have seemed less tolerant of pilots who misuse aircraft than they have been of ordinary citizens who misuse chain saws, power lawn mowers, farm equipment, and similar products to which more people are exposed. It could be that the courts realize that learning to fly takes a good deal of training and education, which would necessarily include proper operation of the airplane by the pilot, and the recognition of areas of misuse that could cause him to come to grief. In like thinking, the courts realize that the average consumer is not particularly well schooled in the finer arts of chain sawing and the operation of other such power equipment; therefore, misuse is more likely to occur, and manufacturers of such products are required to design greater safeguards against them.

The manufacturer must also actually build the product without negligence. If a production worker breaches the standard of care and assembles a component or assembly incorrectly, the airplane is negligently manufactured. Likewise, it is also improperly (and therefore negligently) manufactured if subcomponents that themselves suffer from negligence in the manufacturing process are introduced into, and assembled as a part of, the final product.

With modern manufacturing techniques and proper inspection, there has been a great reduction over the years in the instances of negligent manufacture of not only airplanes, but other manufactured products.

In the final analysis, *a product will be considered by the courts to be defective if it is either negligently designed or negligently manufactured.* It is this key concept of a product's being defective which underlies the heart of product-liability law. *The concept of product liability gives compensation to those who suffer damages as a result of exposure to defective products; negligence in design or manufacture is only one of the ways in which a product may be considered defective.*

WARRANTIES AND REPRESENTATIONS

Very few products are sold on the market today without some type of warranty, guarantee, or representation concerning varying qualities of that product. One of the foundational questions that must be considered is whether a statement concerning the product, its performance, or any other characteristics of it constitutes a warranty on the part of the manufacturer or other supplier; or whether that statement is simply sales talk, and gives no rights to the consumer to expect the statement to be true or to expect the product to perform in accordance with it.

This issue revolves around whether a statement by the manufacturer will be held, by the law, to constitute an *expressed warranty*. There are basically two types of warranties: those that are expressed in words and those that are implied in the law.

An expressed warranty is just that—a warranty that is expressed in words, which can be either oral or written. Most of us are accustomed to dealing with expressed warranties on almost all goods purchased. The typical limited-warranty language that is seen accompanying a toaster, a bicycle, or an automobile is a form of an expressed warranty. The manufacturer is stating in words what it is guaranteeing the product to be, what it is guaranteeing it to do, and how long the customer can expect this particular device to give trouble-free use without expense to the customer for repairs or overhauls.

In some situations, expressed warranties will also address performance characteristics of a particular product. It is in this area that a careful line needs to be drawn between the types of statements that the law will hold to be warranties and those that will be considered sales talk.

Every sales agent who sells the product is entitled under the law to use his best efforts, consistent with not misleading his customer, to sell his product. He is allowed to make claims of opinion, which may or may not be supported with fact. I could easily say that my airplane is one of the best of its kind in the midwest—who is to factually dispute that? That is not a statement of guarantee or warranty, but simply one of my personal opinion. However, when I say that the useful load of the airplane is a certain number of pounds, it cruises at a certain speed, and at that speed it burns a particular amount of fuel, I am stating fact rather than opinion. This is the big difference between warranty and sales talk. When facts are expressed, the law will generally hold those statements of fact to be warranties. Statements of opinion are just that, and do not usually rise to the level of any legally recognizable guarantee or warranty.

When a manufacturer gives an expressed warranty of performance of a product, the law will generally hold that manufacturer to such statements. In the area of aviation law, litigation is frequently associated with the performance characteristics of aircraft. These suits take two forms.

The first class of such actions involves situations wherein consumers are disappointed with the product they purchased, and are claiming that the manufacturer has breached the warranty in the sense that the airplane, helicopter, or other aircraft is not performing as it was represented to perform, or perhaps not in accordance with the performance specifications set forth in the approved flight manual or the pilot operating handbook. These claims are seldom significant, and not frequently successful for the consumer.

Most manufacturers' performance data are extremely well calculated, with proper scientific test flying done to arrive at the statistics that appear in the performance sections of approved flight manuals and operating handbooks. It is a rare airplane indeed that differs from the performance specifications in those documents to any appreciable degree. As we all know, no two airplanes of the same type fly precisely alike, and the manufacturer is allowed a certain degree of leeway. If the stated cruise speed of a specific airplane under certain conditions is 160 knots, and a pilot were to prove in a court of law that her specimen, under those same circumstances, consistently cruises at 156 knots, I seriously doubt that she would be successful in any litigation on the basis of that slight variance from the stated performance.

Where is the dividing line? No one knows for sure, because in the law there are no exacts—only lines which, once crossed, constitute breach of warranty. If the stated cruise speed under the same conditions is again assumed to be 160 knots, and the particular specimen of the aircraft, properly flown under those stated conditions, cruises at only 130 knots, I would have no hesitancy in opining that that aircraft is defective, in that it fails to comply with the expressed warranty of performance under the stated conditions. As previously said, though, 156 knots probably does not constitute a breach. Where is the dividing line? No one knows until a case is tried and a jury is empaneled to decide the *question of fact*: "Is the product conforming to the warranty representations made about its performance?" The jury will decide that issue, and no one can accurately foretell where a group of individuals, in their collective judgment as announced in a verdict, will draw the line between substantial conformance to representations made and breach of warranty.

The other class of lawsuit relating to performance characteristics is a lawsuit that results from an accident, and it is alleged by the pilot or passenger, or the estate of either, that some failure of the aircraft to perform in accordance with the representations made by the manufacturer was the proximate cause of the accident. This type of case generally involves larger risks for the manufacturer than the less complicated consumer action mentioned above.

It is quite possible that performance deficits can cause accidents; in fact, they have. Modern airplanes with the newer-style pilot operating handbook (POH), which is in the format approved by the General Aviation Manufacturers Association, do not suffer from this possibility to nearly the degree as do older aircraft. The modern POH information is extremely complete and well laid out, and leaves very little to guesswork or the imagination of the pilot.

The manufacturer is in a much better position today to defend such a claim concerning this more recently manufactured airplane with the improved pilot operating handbook. If the manufacturer can convince the jury that its information is properly displayed, and that the pilot was negligent in failing to understand the information set forth in the approved flight manual or POH, the manufacturer should be in fairly good stead to defend itself against a large verdict in such a case.

As far as the pilot's culpability is concerned, the reader is referred to the section of Chapter 8 that discusses the applicability of negligence law to pilots. The situation just mentioned is another example of a possible negligent course of action by a pilot.

Naturally, there are controversies that arise concerning the expressed warranties made by manufacturers about the length of time an aircraft is covered for repairs or other breakdowns, and precisely what it is those types of warranties cover, and for how long.

Aircraft manufacturers warrant their new products in a manner very similar to the way in which automobile manufacturers warrant new cars. These warranties are known as *limited warranties*, in the sense that they do not guarantee the entire product to be free of defects in materials and workmanship forever, but make such a statement that is binding upon the manufacturer for a stated period of time. In the automobile business, these limits are usually stated in a specific number of accumulated miles, or span of time, whichever elapses first. In the aviation industry, a very similar method is used, as the manufacturers limit the applicability of their limited warranties to a specific number of flight hours, or span of time, again whichever comes first.

A manufacturer can, of course, face significant liability if a certain airplane, especially an expensive one, suffers premature failure. If the failure is catastrophic, the manufacturer could well be on the hook for a large amount of money, such as in a case where a turbine engine fails within the period of applicability of the limited warranty, and the manufacturer must stand good for a replacement of that engine.

However, in the greater scheme of things, these cases do not usually carry serious exposure to great amounts of liability, because losses of this type are somewhat fixed. The amount it takes to repair the engine, as in the above case, is a finite number, and can be easily calculated. Emotion does not come into play as it does in cases involving death or the pain and suffering of serious injury.

In addition to the warranties that are expressed by the manufacturer, the law will imply certain warranties as well. *Implied warranties* are just that—warranties that the law implies from the fact that certain circumstances have occurred in the business transaction between the parties, and such implied warranties exist without the necessity of either party's having true knowledge of that implied warranty until such time as a claim for breach of it arises.

One implied warranty is the *implied warranty of merchantability*. This warranty stems from the Uniform Commercial Code, which is a set of uniform laws that have been adopted in almost all the states. In essence, it requires that a particular specimen of a product be of sufficient quality that it will pass in the trade as an acceptable example of what it is. In other words, to pass the warranty of merchantability, a product simply needs to be an acceptable specimen of its type and not, of itself, grossly deviant from the others of its type.

To use our example of cruising speed again, if a certain Flythrasher is supposed to cruise at 160 knots and cruises only at 130, one should not have too much difficulty in assuming that that particular specimen is not merchantable, and would not pass in the trade as an acceptable example of the Flythrasher type. But if it cruises at only 156 knots, it probably would pass.

Again, there is no definitive dividing line that can be predicted in advance; whether an example of a product is of such quality that it passes as an acceptable specimen is a question of fact for a jury to determine in a courtroom at the conclusion of a trial. Very few of us will engage in profound speculation of what juries are going to do. Obviously, most lawyers have opinions regarding what a jury might do when confronted with a certain situation, but opinions are opinions, not fact. So the point at which a jury will decide that a particular machine does not pass in the trade as an acceptable example of its type will not be known with certainty until the verdict is rendered at the conclusion of the trial.

The other implied warranty is the *implied warranty of fitness* for a particular purpose. This warranty does not arise nearly as often as the implied warranty of merchantability, nor is it as frequent as the various and sundry expressed warranties that manufacturers might make concerning their products.

The implied warranty of fitness for a particular purpose arises in a limited factual circumstance. That is, before the implied warranty of fitness for a particular purpose can be found to exist, the buyer of a product must make known to the manufacturer what particular purpose is sought, and the manufacturer must agree, in principle, that its product fits that purpose.

Examples illustrate this concept. Assume that an aviation operator approached a manufacturer with the following request. The operator made known to the manufacturer that specific needs included being able to transport a number of workers, up to four, between two mountain airports, both of which have relatively short runways of 2000 feet, and sit at an elevation of 5000 feet above sea level. These airports are both in the western United States, where summertime temperatures will approach 95 to 100 degrees Fahrenheit in the middle of the day, and density altitudes involved in takeoffs under those situations will obviously be extremely high, possibly as much as 10,000 feet or more. Furthermore, this operator informs the manufacturer of the range between the two airports, and the weight of the average load of workers and the equipment that needs to be transported between those two airports.

If the manufacturer then recommends one type of airplane available for sale as a type that will meet those needs and fly that mission, that manufacturer has then stated that one of its products is fit for a particular purpose. Under our scenario no written warranty exists stating that the airplane will be able to fly out of those fields at those density altitudes with the specified load. However, the manufacturer, having knowledge of the unique needs of this particular operator, has in essence impliedly guaranteed, by recommending the particular model to the consumer, that its airplane will perform that function. This is a classic example of the situation that later could be held by a court to give rise to an implied warranty of fitness for a particular purpose.

If, in the course of trying to fly these stated missions, the consumer finds that the airplane is incapable of doing what it was purchased to do, he would have a good chance of success in a lawsuit against the manufacturer, alleging that the manufacturer has breached its implied warranty of fitness for a particular purpose. If a buyer has a particular need in mind, he should make his desires known to the seller of the aircraft under consideration.

An implied warranty of either merchantability or fitness for a particular purpose can be disclaimed by the manufacturer if appropriate wording is used. But there is certainly no harm to the buyer in making sure that particular requirements are made known, so that she can take advantage of the existence of an implied warranty of fitness for a particular purpose in the event that the warranty's existence is not properly disclaimed.

Likewise, manufacturers and sellers of aircraft need to be wary. Each needs good legal advice, if an implied warranty of merchantability or fitness for a particular purpose is not desired in a particular transaction. Again, these warranties can be disclaimed, and their operation avoided, if done properly. If not, and if the circumstances are appropriate, the implied warranties will arise regardless of the desires of the seller, and perhaps regardless of direct communications made by the seller that the implied warranties are not applicable.

Warranties are both a shield and a sword. Expressed warranties, when properly constructed, can limit a manufacturer's obligations, as well as serve as a marketing tool in the manufacturer's sales program. They also let the buyer know precisely how the manufacturer stands behind his product, to what degree, and for how long. Implied warranties can create more difficulties for the uniformed or unwary because, by their very nature, being implied, their terms and applicability are not out in the open for all to see and consider during the negotiation of the purchase of the product.

Those forewarned about the existence of implied warranties, and others who carefully review the terms of expressed warranties, are in a much better position to evaluate the proposed purchase.

Lastly, but certainly not to be ignored, keep in mind that not all representations made by manufacturers or sellers of products will rise to the level of expressed warranties. The old adage "Get it in writing" does not always apply, but the courts are much more likely to hold that a statement concerning a fact is an expressed warranty than they will give credence to statements of opinion about some aspect of a product, especially value.

Value is almost a matter of opinion. If I state that my aircraft is one of the best of its type, and that if a prospective purchaser continues to take good care of it, as I have, it should bring him top dollar when he chooses to sell it, I have made a statement of opinion, certainly not one of fact. Too many consumers are not attuned to the difference between fact and opinion. Opinions are fine, and reputable salespeople offer them all the time. Under the proper conditions, opinions can properly guide the consumer in making a decision concerning the purchase of a product. But all buyers and sellers must be aware that opinions are not warranties, and they do not arise to the level of statements of fact.

STRICT LIABILITY

The issues of manufacturer's liability and negligence discussed above relate to a system of liability that is based upon fault. Most of the American legal system is based upon the ability of one person, who is either free of fault or guilty of fault to a lesser degree, to recover from the person whose fault caused either all the damage in question or at least the majority of it.

Strict liability in tort is a relatively new concept, having been developed during the twentieth century, which holds that *a manufacturer can be liable for losses associated with its product, even though no particular fault on the part of that manufacturer can be proven by the plaintiff in the lawsuit in question.* It is this strict liability that has caused most of the debate about the entire product-liability area of the law.

Those who represent and speak for the interests of plaintiffs will generally postulate that a manufacturer which builds a defective product and places it in the stream of commerce ought to be answerable for any damages caused by that defective product, regardless of whether a plaintiff can successfully prove any particular aspect of fault on the part of the manufacturer.

Those who speak for manufacturing interests naturally feel to the contrary: that a person ought to be responsible if he does not live up to the standard of care applicable to a particular enterprise, but ought not to bear liability when fault is not present or, at the least, cannot be shown to a jury by the preponderance of the evidence that the law requires.

As in all such debates, and as particularly apparent when considering the entire issue of product liability, there is no simple or foolproof solution to this problematic query. Leave it to say at this point that the law of strict liability is well ensconced in the American legal system, and is implied in virtually all states. We must leave the philosophical debate to those who profit from that endeavor; the purpose here is to acquaint the reader with the present situation and possible changes that might occur in the reasonably near future.

The essence of the concept of strict liability is that the plaintiff must prove only that the product is defective. Once the plaintiff carries this burden of proof, there are fewer defenses available to the manufacturer than can be used in a case based upon another theory such as negligence in manufacture and design. For in a negligence case, the manufacturer can also claim that the plaintiff was contributorily negligent and either bar the plaintiff's recovery or reduce it to some extent. If the reader is confused in this area, a review of Chapter 7, specifically the areas concerning defenses to negligence, is in order.

In a strict liability case, those defenses that are available in a negligence case are not all useful to the defendant manufacturer. For instance, the manufacturer cannot claim that the pilot was contributorily negligent, but it can still attempt to show that the pilot assumed the risk of the use of the defective product if it can prove that the pilot had knowledge of the defect and went ahead with his flight regardless of that foreknowledge.

There are several ways in which a plaintiff can strive to show that a product is defective. The first and most obvious is to point toward a discernible and measurable mechanical defect in the product. There have been many cases in the strict liability field that surround mechanical failure of the product or the component parts of it.

Much has been written in the last decade or so concerning the abysmal failure rate associated with dry vacuum pumps and the often catastrophic effects if that failure occurs during actual IFR flight. Although most instrument pilots are originally trained and tested to be able to fly the airplane by reference to nonvacuum instruments, very few are continually proficient in the art of partial-panel flying. I dare to say that very few students, in the course of obtaining an instrument rating, are taught to actually fly an ILS approach down to minimums without use of either the vacuum-driven directional gyro or attitude indicator.

When the vacuum system fails—and that failure is usually associated with a catastrophic destruction of the pump that produces the vacuum to operate those instruments—only the highly proficient and competent instrument pilot will be able to successfully complete the flight in clouds through an approach to minimums. If the failure occurs in meteorological conditions that permit the aircraft to descend into VFR conditions, most instrument pilots would probably be able to conduct a straight descent and get out of trouble.

As is well known to those of us who have suffered vacuum failures, the insidious thing is that, unless the aircraft is equipped with some type of low vacuum warning system other than and beyond the indication of the suction or pressure gauge, the slowly failing vacuum-driven instruments might lead the pilot astray, even into very dangerous situations before she becomes aware that she has, in fact, suffered a vacuum failure.

We have all seen, or read about, the results of this type of failure. After the initial trauma and shock to the families of those who are most likely severely injured or killed if the failure occurs during hard IFR operations, it is very likely that the manufacturer of the airplane, and most probably the manufacturer of the vacuum pump, will be faced with a lawsuit alleging strictly liability for the defect in the system that allowed the failure to occur.

In this type of case, the plaintiff will attempt to prove that the system was defective, and will not actually be required to get into the finer details of attempting to prove that the vacuum system was negligently designed or negligently manufactured. The simple fact that it failed will be the thrust used by the plaintiff to show that the system, or a component part of it (the pump), was defective, and that defect gives rise to a strict liability cause of action for the injuries or deaths that resulted from whatever accident ensued after the vacuum system failure.

This type of case might involve questions beyond the defective character of the vacuum system. Perhaps the plaintiff will go on with his proof to show that the manufacturer further constructed, and placed into the stream of commerce, a product that is defective because there are known methods and systems available to provide a backup to a mechanically driven vacuum pump. Perhaps the same plaintiff will endeavor to convince the jury that the product was defective because there was not an enunciator system installed on the aircraft to warn the pilot immediately that the vacuum pump had failed; hence the product failed to give the pilot the necessary warning that a dangerous condition had manifested itself in time for him to avoid being lured on by instrument indications that were faulty.

The example ought to be clear at this point. The plaintiff needs only to prove that the product was defective, and that such a defect was the proximate cause of the harm that resulted. This is but one vehicle available to the plaintiff, in a strict product-liability action, to show that the product was defective. There are others, and in most situations, a skillful plaintiff's attorney will seek to establish liability on the part of the manufacture alternatively, by alleging and introducing evidence that the product was defective under more than one of the available legal theories that can place legal responsibility for the accident at the feet of the manufacturer.

THE CONSUMER EXPECTATION TEST

Courts have held that a product can be considered *defective in either design or formulation when it is more dangerous than an ordinary consumer would expect, when used in an intended or reasonably foreseeable manner.* It is plain to see that, involved in this test, are elements that otherwise could be considered as negligence, and to some degree, a consideration of the manufacturer's duty to foresee reasonable misuse.

The consumer expectation test has been sorely criticized by many manufacturers and other defense interests. The basic problem with the use of this test, as seen by the defense-oriented interest groups, is that it allows the jury a great deal of latitude to consider evidence that is certainly less than precise. What does a normal, ordinary consumer expect from the particular product? This question is wide open and does permit the jury to weigh the evidence that might or might not be derived with any precision.

The jury must consider whatever dangerous propensities exist about the particular product in question. Evidence must be presented by both the plaintiff and the defendant concerning whether the product is dangerous, and if so, how dangerous it is. Then, we must address what is always a theoretical question concerning how dangerous an ordinary consumer of that product would expect it to be.

If the jury concludes that the specimen which failed, in some manner or another, and about which the suit is pending is more dangerous than an ordinary consumer would expect an ordinary specimen of that product to be, the jury will then hold that the product is defective.

The jury might not even be required to inquire as to whether the specimen that failed and caused the lawsuit differs in any respect from its kindred. If the jury decides that the product in question is more dangerous than an ordinary consumer would expect it to be, when used in either an intended or reasonably foreseeable manner, the product will be held to be defective.

Quite frequently in product-liability suits, the defendant tries to defend by showing that the use to which the product was put was not a reasonably foreseeable use. Again, remember that a reasonably foreseeable use includes reasonably foreseeable misuse. The manufacturer is given considerable latitude to go into the subject of what the intended use of the product is and how it went about reasonably foreseeing other manners of use. Undoubtedly, the manufacturer will attempt to convince the jury that the use to which the product was put, which use resulted in the accident in question, was neither intended use nor reasonably foreseeable use of the product. If the manufacturer is successful in this endeavor, the jury might hold that even though the product is dangerous, it has not failed the consumer expectation test because it was not used in either an intended manner or in some other reasonably foreseeable manner.

For the foregoing reasons, the consumer expectation test has come under significant attack by many legal scholars, but it is becoming a part of the product-liability law in most states. With the general liberality of the law in this area, and as product-liability theories continue to develop, the consumer expectation test will probably be with us for the foreseeable future.

THE RISK-BENEFIT COMPARISON

Another legal theory under which a manufacturer can be held liable for the damages that result from a defective product is the risk-benefit comparison. This theory is one that deems a product defective in either design or formulation when, as it left the control of the manufacturer, the foreseeable risks associated with its design or formulation exceeded the benefits associated with that design or formulation. This test is completed by balancing the foreseeable risks associated with a particular design or formulation of a product against the benefits associated with it.

The law has many balancing tests, and this is certainly not a unique one. The risk-benefit comparison basically takes into account what risks are inherent in the product and what benefits result from its existence. If the risks outweigh the benefits, the manufacturer might be liable. If the benefits are judged to outweigh the risks, the manufacturer is usually able to avoid liability.

Several risks have to be taken into consideration. They include the nature and magnitude of those risks of harm that are associated with the design or formulation of the product when seen in light of the intended and reasonably foreseeable uses, modifications, or alterations of that product.

Also, the likelihood that the design or formulation will cause harm in light of those intended and reasonably foreseeable uses, modifications, or alterations of the product must be viewed by the jury. Like all other questions of fact, the jury will decide, in the ultimate outcome, whether the risks outweigh the benefits, or whether the converse is true.

Evidence probably will be introduced concerning the likelihood of a degree of awareness by the users of that product regarding the risks of harm. This awareness can come about through warnings given by the manufacturer, general knowledge in the industry of the risks of a particular product, or common knowledge among the population as a whole. The law is not particularly concerned with the source of this awareness of product use and associated risks, but simply with the fact that the awareness exists.

As far as aviation products are concerned, one of the most important elements of the risk-benefit comparison is the extent to which the design or formulation of the product conformed to any applicable public or private product standards that were in effect when the product left the control of its manufacturers. Obviously the FAA, in promulgating the sections of the regulations that govern airworthiness and certification of aircraft and component parts, has published a set of public standards that control many aspects of the manufacturing of those final assemblies and their component parts.

Manufacturers often defend product-liability actions on the basis that the aircraft met all the applicable certification rules and that an independent body, namely the Certification Branch of the FAA, had ample opportunity to judge the aircraft and determine whether it met those standards. If the aircraft does not meet the standards, it will not in most instances be certified, and the question would only be theoretical.

Because most aviation product-liability lawsuits involve products that have been certified, this area of the law presents a ripe field for defense efforts on the part of the manufacturers. However, simply meeting the certification standards and getting the aircraft certified will not, of itself, be an ironclad shield for the manufacturer. As with many areas of the law, this is one where there is no absolute defense that precludes success on the part of the plaintiff.

Another recourse used by manufacturers in defending aviation product-liability suits is the doctrine of preemption. This legal principle arises when the federal government properly moves into an area of regulation, and the resulting federal law is in conflict with preexisting state law. The United States Constitution provides that federal law is superior to state law, and when the two conflict, the federal law will control, since the federal government has preempted that particular field.

Since the federal government, through the FAA, has promulgated sections of the FAR that regulate the certification standards for aircraft, inventive defense lawyers have argued that the certification regulations set the exclusive federal standard for the safety of aircraft, and that any notions to the contrary, contained in state law, must yield to the supreme federal law. Unfortunately for the defense interests, this argument has not met with much success. Courts have held that the FARs set only a minimum standard for certification of aircraft, and Congress has not intended them to be the final word on what is, or isn't safe.

This theory is simply one of a number of obstacles that the manufacturer can use to attempt to prevent the plaintiff from successfully completing all the necessary steps toward a favorable plaintiff's verdict.

After looking at the foregoing risks, the manufacturer undoubtedly will introduce evidence concerning the benefits of the challenged design. These benefits must be weighed and compared with the risks that have just been listed.

The first of the benefits to be shown to the jury revolves around the intended or actual utility of the product, including any performance or safety advantages associated with that particular design or formulation. In the aviation field, manufacturers often tout their products as having a higher degree of utility, including performance, than do similar products. If the manufacturer can successfully convince the jury that this particular product has safety and performance advantages that are superior to those of the competition, it will go a long way toward showing that the intended or actual utility of the product is a benefit that must be considered to outweigh the risks that have been propounded by the plaintiff.

Another significant benefit issue that the jury must consider is the technical and economic feasibility of using an alternative design or formulation for the product in question.This query is more commonly known as the *state-of-the-art question*. As fast as modern science and engineering are advancing, aircraft are in the fleet today that were designed and manufactured in a time when technical and economic feasibilities dictated a different design or formulation than if the aircraft were being designed from the ground up at a later time. The jury should receive a significant amount of evidence concerning the state of the engineering and manufacturing art at the time the particular aircraft was engineered and designed. This state-of-the-art defense, while theoretically quite useful to manufacturers, often gets lost in the massive amounts of information that a jury must assimilate in the course of a prolonged product-liability trial.

Nonetheless, when competently and properly presented, the issue of whether the technical and economic feasibility at the time of design and construction allowed an alternative design is one that can be used to offer the manufacturer a meaningful defense.

The nature and magnitude of any foreseeable risk associated with such an alternative design or formulation is an element of the benefits of the particular design or product in question. This element touches also upon the state of the art question. As technology advances, risks associated with new products usually decline. There is a time when manufacturers can introduce new designs or manufacturing techniques when they are reasonably safe; and to have introduced them earlier would not have been a benefit to the particular design.

An example of this area of thinking is the rapidly increasing use of composite materials in aircraft construction, materials which are coming into play to replace structures that were previously made of either wood or metal. Just a few years ago, composite materials (in nonexperimental aircraft) were relegated to nonstructural parts and used in fairings, wing tips, engine cowlings, and similar areas of the aircraft that are not generally subject to high structural loading, and where failure of the particular component would not result in catastrophic failure of the basic structure.

But today, we are seeing all-composite aircraft where the only significant use of traditional metal in the manufacturing process is in the engines and landing gear. There is even talk of future engines to be built of ceramic and other nonmetallic material. There was certainly a time in the not-too-distant past when the nature and magnitude of the risk associated with these new manufacturing techniques would have been too great to allow such an aircraft to be placed in the ordinary stream of commerce.

Thus, as technology marches on, manufacturers must be constantly aware of the foreseeable risks associated with alternative designs or formulations; and when the plaintiff alleges that some other material, design, or manufacturing technique should have been used to create a safer product than the one at issue, the manufacturer might well be in a position to point to the large risks associated with the new techniques and defend the continued use of more traditional approaches to the design or manufacture of the challenged product.

Likewise, a product will not be deemed to be defective when the harm caused by it was the result of an inherent characteristic in that product that is a generic aspect of it that cannot be eliminated without substantially compromising the product's usefulness or desirability, and that is recognized by the ordinary consumer with the ordinary knowledge common to that industry. That defense is certainly a mouthful, and contains many considerations that the jury must weigh.

Virtually all products have some inherent risk or danger. Aircraft will stall if not flown at a speed at which the wings can continue to produce effective lift. Engines will fail if not operated properly. There are many other generic risks associated with aviation products that cannot be eliminated without substantially compromising the product's usefulness or desirability. A good example of this situation is that normal airplanes will spin if conditions are conducive at the moment they undergo an aerodynamic stall. There have been attempts over the years to build spin-proof airplanes, and at least one aircraft claiming to be spin-proof was certified in the 1940s. But at a time when many other manufacturers were seeing tremendous economic success in their business, the manufacturer of that particular product was unable to sell enough units to remain in business. There were problems with the airplane associated with the limited up-elevator travel that was required, together with limited rudder travel that was built into the aircraft, to prevent the stall-spin combination from occurring.

The marketplace determined that there was a substantial enough compromise of that particular aircraft's usefulness or desirability that it did not survive the acid test of sales.

Another part of this element of benefit is that the ordinary consumer, with ordinary knowledge of the industry, must know of this generic and inherent characteristic of the product. All pilots should know that airplanes can stall, and even though modern airfoil designs have, to a great degree, lessened adverse stall characteristics, most fixed-wing aircraft are capable of being spun if conditions, most of which are induced by the pilot, are right. The manufacturer must show that these generic characteristics, applicable to all airplanes, are not any more risky in the aircraft challenged in the case than in airplanes in general. If the manufacturer is able to do this, it goes a long way toward a successful defense.

In addition to that, the manufacturer must show that at the time the product left the manufacturer's control, a practical and technically feasible alternative design or formulation was not available that would have prevented the harm for which the plaintiff seeks damages without substantially impairing the usefulness or intended purpose of the product. If the manufacturer could show that these considerations were present, and that no practical or technically feasible alternative design or formulation would have resulted in a product that could have prevented the accident, the manufacturer would again put up an obstacle that the plaintiff might have difficulty crossing.

As a side comment, the ability of the manufacturer to defend on the basis of the lack of a practical or technically feasible alternative will not be available if the jury finds that the manufacturer acted unreasonably in introducing the product into trade or commerce. A manufacturer might well act unreasonably in manufacturing the particular aircraft if, even though there is no practical and technically feasible alternative, the state of the art that is available resulted in a product so dangerous that no manufacturer would reasonably introduce it into the aviation community.

Therefore, the risk-benefit comparison is probably one of the most difficult tests in the product-liability field for the jury to comprehend. There are so many factors at work that the jurors, none of whom are likely to be pilots themselves or otherwise meaningfully educated or experienced in the aviation industry, will have a tremendously difficult time making the comparison that is required between the risks and benefits of the product in question. This is just another example of an area where the legal system is not perfect, and probably never will be. But with competent and able legal representation, the parties should be able to distill the various claims that each wishes to present to make them few and simple enough that the collective mind of the jury can come to a just decision.

FAILURE TO WARN

Manufacturers have long been held to the duty to warn their consumers of the correct uses of a given product and risks associated with it.

In the aviation world, manufacturers have progressed greatly in the last two to three decades in providing effective warnings and instructions concerning the use of airplanes. It takes only a few moments to compare the pilot operating handbooks of the 1950s and 1960s with the modern GAMA format in use today. A good example is a momentary look at those earlier manuals and the paucity of information they contained. Many of them would specify only an airspeed for best rate of climb, without mentioning a speed for best angle of climb. The performance charts would seldom give adequate instructions for applying correction factors to determine the effects of density altitude upon takeoff and climb performance.

In this area, the modern-format pilot operating handbook for a light single-engine airplane equipped with a constant-speed propeller and retractable landing gear would be comparable to the type of handbook for a large multiengine airplane produced just three or four decades ago.

Many manufacturers could increase their ability to defend product-liability cases on the basis of the lack of effective warnings by updating the pilot operating handbooks for many of the older aircraft that are in significant use in the general aviation fleet today.

The law requires that at the time a product leaves the control of the manufacturer and the manufacturer knows or, in the exercise of reasonable care, should know, about a risk associated with that product, the manufacturer must provide warnings or instructions that a manufacturer exercising reasonable care would have provided concerning that risk. Further, this requirement must be seen in light of the likelihood that the product could cause harm of the type for which the plaintiff in the lawsuit is seeking to recover, and the likely seriousness of that harm. This is a difficult test, and it is one that raises its head in virtually every product-liability case.

One area in which manufacturers of older aircraft are able to defend allegations of inadequate warnings or instructions is the portion of the legal requirements that relates to the reasonable care that a manufacturer would have provided concerning the warning or instruction. Again, an example is the older pilot handbook. In a time when scanty information was the industry norm, most manufacturers will defend inadequate warning cases concerning older aircraft on the basis that they gave the same kind and degree of warnings and instructions that were used by everyone else in the industry at that time, and therefore did what was reasonably required in the exercise of reasonable care in terms of warning and instructing pilots of those airplanes.

Juries often get confused in this area and hold manufacturers to a fairly strict standard of accountability. Because aviation was a reasonably well-developed science in the 1950s and 1960s, there is very little that can be done to convince jurors that manufacturers were unable to provide the detailed instructions and warnings that are commonplace today.

After a product has been placed in the market, further warnings and instructions might be required of the manufacturer. A product will be deemed to be defective if, at a relevant time after it left the control of the manufacturer, the manufacturer knew about, or in the exercise of reasonable care should have known about, a risk associated with the product that allegedly caused the damages sought, but failed to provide the postmarketing warning or instruction that a manufacturer would in the exercise of reasonable care. Once again, this requirement must be seen in light of the likelihood that the product would cause harm of the type for which the plaintiff is seeking to recover, and also in light of the likely seriousness of that harm.

It is this legal requirement, together with good business sense, that makes manufacturers issue *service bulletins* and other postmarketing instructions. When an aircraft manufacturer becomes aware of some aspect of the use of its product that requires a warning or instruction, it is now required under the law of most states that a communication be issued to the users of the airplane, if they can reasonably be located, and to the community at large, about that newly discovered problem.

In recent years, manufacturers have taken advantage of the ability to defend themselves from failure-to-warn cases by issuing a proliferation of postmarketing warnings and instructions concerning the use of certain products. Because there is no amount of testing that can completely duplicate the effects of thousands, if not hundreds of thousands, of hours of use in everyday flying, manufacturers routinely become aware of situations revolving about their product that require some warning or instruction long after the aircraft is placed into service.

A good many airworthiness directives that are issued by the FAA come about as a result of the efforts of manufacturers to put postmarketing warnings, alterations, or fixes to their aircraft into effect. Many manufacturers have been held liable for defective conditions in their products even after they attempted to adequately warn the consuming public of those problems. But if a manufacturer is able to convince the FAA to issue an *airworthiness directive*, then the situation must be resolved, and it is much more likely that the manufacturer can build a successful defense around the failure of the aircraft operator or pilot to comply with the regulatory requirement for the mandated alteration or repair.

Also, a product will not be considered to be defective because of lack of warning or instruction if the risk that caused the harm for which the plaintiff is seeking recovery is an open and obvious risk, or if it is a risk that is a matter of common knowledge. Manufacturers can always try to defend with the theory that the product user intelligently and voluntarily exposed himself to an obvious and known risk involved in the use of the product.

We all know, for instance, that propellers are certainly an aspect of the ordinary general aviation airplane that contain a great degree of risk. But the danger of walking into a propeller or sticking one's hand or other body part into a rotating propeller has been deemed to be a matter of such common knowledge as to not require warning or instruction from the manufacturer. Likewise, the failure of the user of an aircraft to properly secure herself within the bodily restraints in the nature of seat belts or shoulder harnesses is so obvious, and known to the common person, that a great deal of instruction and warning need not be given by the manufacturer about that subject.

In conclusion, the failure-to-warn theory is one which is alive and well in product-liability cases. It is an area of the law that is developing rapidly, and manufacturers might well find themselves under an increasing burden to discover and warn or instruct users of the product of the potential risks associated with it and how to best eliminate or reduce the likelihood of serious harm involved.

GENERAL AVIATION REVITALIZATION ACT OF 1994

Without excessively rehashing the material at the beginning of this chapter, the increased exposure to product-liability suits has been cited by many pundits, and manufacturers, as the primary cause for the economic debacle in which the general aviation industry has found itself since the peak production years of the late 1970s. The absolute apex of light plane production, when measured in terms of units manufactured, occurred in 1978. Just 8 years later, in 1986, Cessna rolled its last single-engine piston-powered airplane out of the factory doors. That company resoundingly stated, year after year, that it would not get back into the business of making new single-engine airplanes until something major was done to control product-liability problems.

Cessna's tale of woe is not the only one—all the light plane manufacturers had similar drastic cutbacks in production. In 1995, total piston engine production, again measured in units made, not total dollars of sales, was at a level of only 5 to 6 percent of the 1978 peak. Virtually no other industry has ever seen a 94 to 95 percent drop in business, over so short a period of time, and survived.

The General Aviation Manufacturers Association (GAMA), along with the help of a few selected senators and representatives, launched an effort in the mid-1980s to pass some sort of federal law that would deal with the product-liability issue, and tighten the rules, to the benefit of the manufacturers. So in several sessions of Congress, various bills were introduced. Most died in committee, that unique congressional method of keeping a bill from coming to a floor vote.

Plaintiffs' interests lobbied heavily to kill any such attempt at congressional regulation of the product-liability field, knowing that if one of these laws passed, the doctrine of preemption would likely make the federal law uniform across the country, and severely restrict the rights that plaintiffs enjoyed in a substantial number of states.

Without overpoliticizing the discussion here, the defense interests were finally successful in 1994. The resulting federal law, called the General Aviation Revitalization Act of 1994 (GARA), significantly changed the rules. As soon as the law was signed by the President, the champagne corks popped in Wichita, Vero Beach, Kerrville, and other centers of general aviation manufacturing activity.

The title of the law says what it is intended to do, which is to revitalize the general aviation industry. Since Cessna has once more begun making single-engine, piston-powered airplanes, with initial production targets of 2000 airplanes per year, the law has made a good start at fulfilling its goal. Let's take at look at some of the definitions in this landmark law, and then we'll examine what GARA does for the manufacturer.

The law is applicable to general aviation aircraft, as it defines them. This term is used to describe an aircraft that has a type certificate or airworthiness certificate issued by the FAA. At the time of such issuance, the maximum seating capacity of the aircraft can be no more than 19. At the time of the accident which is the subject matter of a product-liability suit, the aircraft cannot be operating in scheduled, passenger-carrying flights for GARA to be effective to that lawsuit.

GARA goes on to establish what it calls a limitation period, or cutoff period, of 18 years after the occurrence of certain events. If any of the listed events happened more than 18 years before the accident, no one who was aboard the aircraft at the time of the accident can sue a manufacturer for damages resulting from that accident. The 18-year period starts running on:

1. The date of delivery of the aircraft to its first purchaser or lessee, if delivered directly from the manufacturer; or
2. The date of first delivery of the aircraft to a person engaged in the business of selling or leasing such aircraft

So, after 18 years from delivery to the first dealer, purchaser, or lessee, the manufacturer of the basic airframe is off the hook. When you look at the production figures by year of manufacture, keeping in mind that peak production happened in the light plane industry in 1978, it becomes apparent that by 1996, general aviation manufacturers were immune from suit for the vast majority of the aircraft comprising the general aviation fleet.

The federal law also deals with components that are replaced. The same 18-year period of vulnerability applies to the manufacturer of any new component, system, subassembly, or other part which replaced another component, system, subassembly, or other part which was originally in, or added to, the aircraft and which is alleged to have caused such death, injury, or damage. The 18-year limitation period begins on the date of completion of the replacement or addition.

The second part of GARA, concerning replacements or additions, will create a flurry of litigation over what is meant by some of the terms used. The courts will be called upon to answer such questions as what is meant by a new component. When an engine is overhauled, new parts like pistons and valves are almost always used, whereas old parts like crankshafts and crankcases are most often reused.

Does the maker of the new piston get protection from GARA, while the company that made the reused crankshaft is not covered, since that part isn't new? No one will know for years, until cases get filed, tried, and appealed in the justice system.

One thing is for sure, and that is that the basic airframe manufacturer gets a tremendous amount of protection from GARA, once the basic airplane is 18 years old, which almost all of them are. Since engines, alternators, vacuum pumps, and avionics seldom last 18 years, their makers will not see the same benefits. But this statement depends in great measure upon how the courts will deal with the questions of what constitutes replacement. It could well be that the maker of reused parts will be protected, when those parts are reinstalled in an engine during overhaul. We'll have to wait and see how the law is interpreted.

There are a few exceptions to the prohibition against lawsuits after the 18-year limitation period. First, GARA doesn't apply if a passenger is on board an aircraft for the purposes of receiving treatment for a medical or other emergency. So, EMS flights which result in accidents still leave the manufacturer exposed to whatever the state law is, without considering GARA. The law also doesn't apply to anyone injured who was not aboard the aircraft. If a person is in a building, on the ground, or likewise not in the aircraft when the accident happened, that individual can still sue for damages irrespective of GARA.

If a plaintiff can prove that the manufacturer, with respect to a type certificate or airworthiness certificate for, or obligations with respect to continuing airworthiness of the aircraft or any component system, subassembly, or other part, knowingly misrepresented to the FAA, or concealed or withheld from the FAA any required information which is relevant and material to the performance, maintenance, or operation of the aircraft, or any component, subassembly, or other component part, GARA's protection isn't applicable to any lawsuit if that misrepresentation, concealment, or withholding of required information is causally related to the accident.

Lastly, GARA doesn't apply if the manufacturer is sued under some written warranty that would otherwise be enforceable in court. There are not many written warranties that are operable for extended periods of time, so this exception will probably not be all that important.

The most important provision of GARA, besides the 18-year limitation period, is the part of the law that makes it preempt all state laws that would permit a civil lawsuit to be brought after the limitation period. Congress has clearly exercised its power of preemption.

Therefore, even though GARA doesn't go all the way toward requiring that all general aviation product-liability litigation takes place in the federal courts, as some earlier and unsuccessful attempts at this kind of legislation did, it makes the 18-year limitation period uniform throughout the United States. Only time will tell the true effect of GARA in reducing the expense that manufacturers suffer from product-liability cases concerning older aircraft.

The philosophy of the law is simple: Eighteen years is long enough in which to determine if the aircraft is defective. If it flies without accident for that long, the law presumes that it is not defective, and its manufacturer should be shielded from product-liability litigation about it.

SUPPLIERS' LIABILITY

Suppliers of products are defined and treated differently from those who manufacture goods that end up in the stream of commerce. Courts usually define a *supplier* as a person who, in the course of a business that is conducted for that purpose, sells, distributes, leases, prepares, blends, packages, labels, or otherwise participates in the placing of a product into the stream of commerce. In particular situations, the term *supplier* will also include persons who install, repair, or maintain any aspect of a product that allegedly causes harm.

Suppliers do not include those who manufacture products, nor does the term include sellers of real property or providers of professional services who, incidental to the professional transaction, sell or use a product. As can be distilled from the lengthy definition, suppliers are generally those who sell or lease products or who install or maintain them, while not being the manufacturer.

In some situations, suppliers can be held liable in a product-liability claim. The circumstances that engender supplier liability are fewer than, and different from, those that will operate to render a manufacturer liable for a defect in its product.

First of all, a supplier will be deemed liable under a product-liability action if that supplier himself was negligent, and that negligence was the proximate cause of the harm for which the plaintiff seeks to recover damages. In the aviation industry, the definition of supplier, coupled with the circumstance just mentioned, negligence of the supplier, is particularly germane to maintenance facilities. Repair shops that maintain aircraft or component parts certainly meet the definition of suppliers as quoted at the beginning of this section.

When a maintenance facility is itself negligent, that negligence can give rise to a suit under a product-liability theory against it. Naturally, the specific negligence in question must actually be established by the plaintiff to a preponderance of the evidence, and the jury must return a verdict that the repair shop was negligent in the conduct of the work it did on the aircraft. Next, as is the case in all suits based upon negligent conduct, the particular negligence that has been shown to exist must further be the proximate, or legal, cause of the harm that is alleged by the plaintiff.

Thus, if a mechanic negligently repairs and subsequently installs an alternator or generator, but the accident is shown to have been caused by a catastrophic internal failure of the engine itself, for which there is no allegation of negligent repair, the errors committed in the process of working on the electrical system would, in all likelihood, not be deemed to be a proximate cause of the accident, and therefore not of the injuries for which the plaintiff is seeking recovery in court.

Suppliers can put themselves in a position of liability by making representations about products. If the product in question fails to conform to a representation made by the supplier, and that failure and the representation itself were the proximate cause of the damages, the supplier might well be in a position to be successfully sued by a plaintiff who suffered injury as a result.

It should be noted that fraud is not an element of supplier liability in this area. The supplier is still subject to liability for representations and the failure of the product to conform to them, even though the supplier did not act fraudulently, recklessly, or even negligently in making the representation. Those who sell or lease products ought to heed the warning that representations made can become the basis for a product-liability claim, even though there was no fault on the part of the salesperson making the particular statement. Much of the earlier discussion in this chapter concerning situations under which a manufacturer can be held liable for product-liability claims under the warranty theory apply to the supplier as well.

There are certain situations under which a supplier will be subjected to further liability, as if she were a manufacturer. The following circumstances do not occur all that often, but those who fall within the definition of supplier ought to be aware of the times when they can be hauled into court and subjected to the same all-encompassing liability that a manufacturer faces.

First of all, a supplier can be subject to the same liability as a manufacturer if the manufacturer is not subject to what is known as *judicial process* in the state where the suit is filed. This is more of a theoretical than a practical problem in the aviation industry. A liberal approach is used by the United States Supreme Court in defining when a suit might be maintained in a particular state against a business that is headquartered in another state. Therefore, there are not many situations or sets of facts that present themselves today wherein a manufacturer can claim that it is not subject to suit in the state where the action has been filed.

Before agreeing to sell any product, the supplier would do well to check with legal counsel to see if there would be any meaningful problems with the manufacturer's being subject to service of process in the state where the supplier is going to do business. Because if that manufacturer is not so subject to suit, the seller needs to be aware that he might be subject to the broad spectrum of liability as a manufacturer. This warning is particularly germane to the supplier of an aircraft or part which is made by a manufacturer located in a foreign country, and which is not subject to being hauled into American courts.

Next, some states allow a supplier to be subjected to manufacturer's liability if the plaintiff would be unable to collect a judgment against the manufacturer of the product because of the actual or asserted insolvency of the manufacturer. An unanswered question raised in a number of states that permit this type of liability is whether the manufacturer's formal bankruptcy in a United States Bankruptcy Court proceeding results in the actual or asserted insolvency that transfers the manufacturer's liability to the supplier. Again, if the supplier is going to be dealing with products made by a manufacturer of dubious financial standing, that supplier would be well advised to undertake a determination of the law of the state in which most of his products are to be sold and used before making the business decision of whether to actually sell the products of that manufacturer.

The next situation in which a supplier can be subject to manufacturer's liability is when the supplier in question owns, or owned, in whole or in part, the business entity that manufactured the particular product. It is certainly more likely that manufacturers will own subsidiaries to supply their products than it is for the supplier to be the parent company of the manufacturer; but again, an unanswered question of liability can arise when the supplier is an individual who is a casual shareholder of the corporation that manufactures the particular product. There is no clear answer to this query, and if a supplier is going to invest individually in the shares of stock of the company that manufactures the products he sells, he might wish to have this question researched by his own attorneys and obtain advice that enables him to judge the wisdom of either investing in or selling a particular product.

Next, the reverse situation also does apply. If the supplier is owned, or was owned, at the time the product was supplied, either in whole or in part by the manufacturer, that supplier can be held accountable to the same standards as the manufacturer is held. Again, this is the more common situation than the one just preceding.

If a supplier creates or furnishes the design or formulation that the manufacturer used to produce, create, make, construct, assemble, or rebuild a product or a component of the product, that supplier will also face the same degree of liability as the manufacturer. This scenario happens quite frequently when large suppliers dictate to manufacturers the design or formulation of a product to be made for sale by the supplier. Many national retail sales chains frequently dictate to manufacturers the design characteristics of the products to be sold by their stores. When that happens, the supplier has, in fact, become the one who has designed and formulated the product, and the law then has little hesitancy in transferring the liability, which otherwise would rest upon the manufacturer, to the one who did the design and formulation—the selling supplier.

In the aviation industry, some of the larger catalog houses are now offering their own brand of products. It remains to be seen if they participate in the design and formulation stage sufficiently to be deemed by the courts to bear the liability of the manufacturer in a product-liability claim brought concerning those kinds of products.

Suppliers can further be held liable as manufacturers if they alter, modify, or fail to maintain the product after it comes into their possession and before it leaves their possession on its way to the consumer. The modification, alteration, or failure to maintain must render the product defective in order for this theory of liability to be useful by a plaintiff against a supplier; but keep in mind that the definition of defective is so all-encompassing that, if the product does fail, it will more than likely be considered to be defective.

There are many rebuilding shops around the country that are in the business of modifying aircraft through either supplemental type certificates or less drastic forms of modification that simply refine and more finely tune the operating characteristics of the product, as compared with the way the aircraft performed when it left its original manufacturer. Modification facilities definitely must be aware that they are most likely to be held to the liability standards of a manufacturer if any part of their modification or repair process comes into question as the alleged proximate cause of the harm that led to the filing of the lawsuit. This is true whether we are concerned with airframe modifications, engine upgrades, or combinations of the two.

Next to last, a supplier will face a manufacturer's liability if the supplier markets the product under either its own label or its own trade name. This situation is occurring more as frequently the larger mail-order supply houses begin to market products with their own labels or trade names thereon, rather than using the brands normally seen on the product, placed thereon by the manufacturer. If a supplier wishes to represent to the world that a particular product is his, he will bear the liability attendant on his having manufactured it, regardless of whether he did, or whether he has simply purchased that product and attached his own label to it. It makes no difference—the result is the same, and that is the full liability faced by the manufacturer of the product.

Lastly, suppliers have been held to the manufacturer's standards of liability when they fail to respond in a timely and reasonable manner to a request on the part of a plaintiff to disclose the name and address of the manufacturer of the product in question. Long gone from the procedural rules of most states are the concepts that lawsuits will be decided by surprise or ambush. The liberal discovery rules that first saw general acceptance in the early part of this century have become the widespread law of almost all states and, as well, of the federal court system.

If the supplier wants to act as a shield for a manufacturer and, in so doing, deflect claims otherwise addressed to that manufacturer by failing to disclose to plaintiffs who it is who made the product in question, the supplier will be able to do that under some circumstances; but the result will be that the supplier will be held to the same theories of liability as would the manufacturer, if it were identified. Turnabout is fair play, and most suppliers are not going to engage in this type of conduct without suffering the consequences.

In short summary, suppliers can protect themselves from a good number of the product-liability claims brought against them if they do not make representations about products, but rather refer customers to the manufacturer and whatever statements it has made about its goods. Likewise, if suppliers are not negligent in repair and deal with manufacturers that are subject to suit in the states where the products are likely to be sold, such suppliers will have a reasonably good defense to a product-liability claim. The supplier of a product probably will not be subject to product-liability claims if that supplier:

1. Is not negligent in any respect toward the product or its maintenance
2. Does not tell the actual manufacturer how to design or formulate the product
3. Does not engage in significant alteration or modification
4. Does not market products under its own label or trade name
5. Gives truthful answers to inquiries concerning the identity and address of the manufacturer

Since the passage in 1994 of GARA, many aviation lawyers and insurance professionals are predicting that suppliers will become more attractive targets in lawsuits after accidents. Once the traditional deep pocket, the manufacturer, has gained immunity from suit because of the existence of GARA, plaintiffs will have to look elsewhere for their recovery.

Overhaul shops and maintenance facilities are likely to become the bull's-eyes at which plaintiffs aim. Lawsuits won't suddenly go away because of GARA; more likely different folks will be bearing the brunt of the problem.

CRASHWORTHINESS

As a corollary to the legal requirement that a manufacturer reasonably foresees the use and reasonable misuse of its product, the law has, in the last few decades, created a subtheory that the manufacturer has a duty to foresee that its products will, if they are at all subject to accidents, be involved in accidents. The manufacturer then has a duty to use reasonable care to protect product users from harm that results from those accidents.

During the 1960s this theory regarding the duty put on manufacturers was refined into what is now known as the *theory of crashworthiness*. Crashworthiness implies that a product—for our purposes, aircraft—ought to be able to protect its occupants in certain crashes from serious bodily harm or death.

A good bit of the entire subject of crashworthiness involves rather complicated engineering concepts. But boiled down to its essence, crashworthiness implies that the vehicle will protect its occupants when the accident is of such a nature, and contains such forces, that the structure of the aircraft is not so compromised as to prevent an occupant from surviving the accident, and not incurring life-threatening injuries in the accident sequence.

One of the most crucial elements of evidence in a crashworthiness case is whether a livable volume was maintained inside the cabin in which the occupants could survive if the aircraft were properly designed. Without question, there are those accidents that result in such massive deformation of the basic aircraft structure, or even disintegration of the structure, that no amount of occupant protection is going to save the lives of those inside. But to a great degree, aviation accidents are often survivable in the sense that the basic airframe structure is not so deformed or otherwise compromised that the occupants do not have a sufficient space in which to live if the design meets modern concepts of crashworthiness.

First of all, experts will be retained to provide testimony and other evidence regarding the nature of the accident and the forces experienced during the impact sequence. The human body can withstand tremendous amounts of force applied to it, if that force is of sufficiently short duration, and if the person involved is properly restrained.

In an accident involving longitudinal forces, if a person is restrained only by a lap belt, he will flex at the waist and his head, arms, and legs will generally flail to the full amount of extension possible. To get an idea of this concept, picture a person lying on a floor doing a sit-up exercise with arms extended so that when the body is upright and then bent over center, the fingertips come near the person's toes. When the only form of restraint used is a lap belt, it is possible that the body will extend itself fully during the forward impact. Therefore, it is easy to see that significant bodily injury can occur to the upper torso, head and neck, and extremities when the crash occurs.

To the contrary, when a passenger is restrained with proper upper-torso restraints, which are commonly known as shoulder harnesses, the amount of flexion possible in the upper part of the body is drastically reduced. That is the main reason that shoulder harnesses are now required in the front seats of all new fixed-wing aircraft, and, in my opinion, ought to be installed in all passengers seats as well. The ability to restrain the legs is somewhat less practical in general aviation aircraft, but very few deaths result from leg injuries. Far more harm is done when chests and heads impact instrument panels, control wheels or sticks, and windshields, than is ever done by similar striking of those components by feet and legs.

Therefore, manufacturers frequently face product-liability claims concerning the alleged lack of crashworthiness of their products when those aircraft are not equipped with upper-torso restraints. If the cabin collapses around the occupant such that bodily injury or death results from the structural deformation itself, there is not nearly as great a likelihood that a plaintiff will be able to successfully maintain a crashworthiness case. But when the structure remains relatively intact, and there is a sufficient volume of structure within which the occupant could have lived if properly restrained, the manufacturer might face a real problem in a courtroom if the aircraft was not equipped with adequate restraints.

Always, an argument can be made as to the extent of practical considerations in this and other regards. A good number of us realize that the military-style shoulder harness, which comes over both shoulders and meets the lap belt in a common three- to five-point buckle, is far superior to the diagonal, one-strap upper-torso restraint presently used in most general aviation aircraft and also, for that matter, in most automobiles. But manufacturers must be realists. A person is much more likely to use, and therefore gain what protection is available from, a shoulder harness that is more comfortable and that gives at least the perception of minimal interference with normal movement.

Therefore, manufacturers have come to the conclusion that the diagonal, single-strap harness is more likely to be used than the more effective military arrangement. For that reason, there is a system installed in most aircraft that is less than ideal, but still provides a tremendous amount of increased protection over that available with merely a lap belt.

Courts also have held that manufacturers have a duty to delethalize cockpit interiors. This issue was first developed in the arena of automobile accidents, as was a good part of the law of crashworthiness. At least a few of us are old enough to remember when padded dashboards first became available in automobiles in the mid- to late 1950s. It is almost ironic to look at older cars with their sheet metal dashboards that fractured their share of skulls in relatively minor accidents.

Likewise, aircraft interiors have become less dangerous as control wheels have been designed to do less damage and be far less likely to impale the occupant than those in older aircraft. The arrays of knobs, handles, and other control devices often function as knives and spears when the human body is subjected to the forces involved in an aircraft accident. Manufacturers have made great strides in recent years to design and make aircraft interiors, including, but not limited to, the crew positions, far safer in this regard.

The key to the entire concept of crashworthiness is that the volume inside the cabin must remain livable during the crash sequence. If it does, the manufacturer then has the duty to do all that is reasonable to protect the occupants and allow them to survive in that livable space. To do otherwise will probably be held to be a breach of the manufacturer's duty, and will subject it to liability for the damages suffered by the occupants.

Further, the manufacturer can be liable for an enhancement to injuries resulting from the lack of crashworthiness of an aircraft. Assume a situation in which expert testimony is produced and delivered to the jury that shows that the pilot in question suffered a skull fracture as a result of the accident. The same evidence shows that, had that person enjoyed the use of a shoulder harness, the injuries received would have been far less drastic than they were. In such a situation, assuming the manufacturer has no other basis of liability for this accident, it can be held responsible for the increased amount of injury suffered by the occupant over what he otherwise would have received, had the aircraft been considered crashworthy. This is so even if the injured pilot was himself negligent in flying the airplane, and that negligence caused the accident to occur.

Manufacturers are often doing their level best to increase the crashworthiness of aircraft already in the fleet. One of the major general aviation manufacturers has made it clear that it is in the process of designing shoulder harness kits for retrofit into all aircraft currently in its fleet of general aviation airplanes. Even though shoulder harnesses were not in vogue at the time many of these airplanes were designed and manufactured, almost all general aviation aircraft are susceptible to retrofit with reasonably good upper-torso restraint systems. Once the manufacturer makes these kits available, it may be going a long way to absolving itself from liability for the fact that the equipment was not installed in the aircraft when first manufactured.

It is my humble opinion that retrofit shoulder harness kits, besides reducing liability, are a very worthwhile investment for every aircraft owner, and ought to be considered for installation in all the older aircraft that did not come from the factory with them installed.

A manufacturer is held only to the duty to reasonably foresee that their products will be involved in accidents. Therefore, it is a general rule that the manufacturer is not liable, under a theory of crashworthiness, for an unforeseeable accident, or for an accident that occurs in a very unusual or unforeseeable sequence.

A good example of the foregoing is the fact that most helicopter accidents involve high vertical loads and relatively low longitudinal loads. That means that most helicopter accidents involve a rapid vertical descent, with the helicopter remaining in some fashion in a normal skid-level attitude, and impacting the ground in this attitude, but with sufficient force that severe damage can be done to the airframe structure as well as to the persons inside. Unlike fixed-wing aircraft, few helicopters are involved in accidents where they crash at a high forward speed and suffer extreme longitudinal loads. Therefore, the concept of crashworthy designs, as applied to helicopters, is much different than with fixed-wing aircraft.

The crash-attenuating seats found in modern helicopters tend to absorb the vertical impacts, but not a tremendous degree of attention is paid to protecting occupants from the very rare accidents that involve high longitudinal forces that are often seen in airplane accidents. The common physical injury that results from helicopter accidents involves the back and spinal column of the occupants. It is all too unfortunate that some former helicopter pilots are in wheelchairs as a result of the spinal injuries received in helicopter crashes. To the contrary, very few airplane pilots suffer injuries of this type, but generally come to grief by virtue of the high longitudinal forces generated by the higher forward speed at which the airplane usually impacts when involved in an accident.

Manufacturers of all kinds of aircraft have the common duty to make their products as crashworthy as possible, considering the fact that all aspects of aircraft design involve inherent compromises. One other area of design that has improved dramatically in recent decades is the placement of fuel tanks. Those of us who like the classic airplanes of the 1940s and 1950s put up with fuel tanks located right behind the firewall, and directly ahead of the pilot, ready to spew gasoline all over the place in an accident.

Modern regulations for seat coverings and other items of upholstery require them to be fire-resistant, so that toxic gases and flames from the interior are minimized if a fire is a part of the accident's aftermath. The older type of seat belt, which had the belt going through a cam-type buckle, has been outlawed for years, replaced now with a metal-to-metal buckle. Ignition keys in cars and magneto switch keys in light planes now come with a rubberized head, to lessen the chances of injury from striking them.

Therefore, the manufacturer of each particular product has a duty not only to foresee that its aircraft will be involved in accidents, but also to analyze the types of accidents most commonly involving its product and do what is reasonable to protect the occupants from either injury or enhanced injury when a livable volume of space remains available. This area of the law continues to change rapidly, and the last word is far from having been written about product liability.

10

Medical Certificate Appeals and Special Issuance

AS THE RULES NOW STAND, EVERY PILOT OF A POWERED AIRCRAFT, except for a motor glider, is required to pass a physical examination and possess one of the three classes of medical certificates in order to fly. Even student pilots have to have a "medical," as the certificate is commonly called. Medicals come in three classes, and are valid for varying amounts of time.

The lowest class of medical is the third-class certificate. A third-class medical is required of student, recreational, and private pilots. As this book is being written, a proposed change to Part 61 of the FAR is working its way through the rule-making process. If it is adopted as proposed, recreational pilots, or those who have higher-level certificates but who want to exercise only the privileges of recreational pilots, would be allowed to self-certify their physical condition, and would no longer be required to possess a medical certificate to fly powered aircraft appropriate to the recreational license.

A third-class medical is valid for 24 calendar months from the date of its issuance. If you go see an aviation medical examiner (AME) on May 1 of a given year, your third-class medical is good until May 31, 2 years later.

Second-class medicals are required of all commercial pilots and airline transport pilots (ATPs) who are flying for compensation in operations not actually requiring the pilot to hold an ATP. Examples of these types of commercial flying are flight instructing, some charter operations, local sightseeing rides, banner towing, crop dusting, and similar flying activities in which the pilot is paid, but not flying in airline operations. Second-class medicals are valid for 12 calendar months from the date of issuance for commercial operations. After the year passes, the certificate reverts to a third-class medical for the remainder of the 24 months for which a third-class is valid, assuming that the pilot doesn't actually need a second-class medical.

The most strict physical exam is given for a first-class medical, generally required of airline pilots. This medical is valid for only 6 calendar months from the date of issuance. After that time, it reverts to a second-class medical for 6 more months, then to a third-class certificate until 24 calendar months have passed since it was issued. As a first- or second-class medical reverts to a lower class, it is valid for only those pilot operations requiring the lower-class certificate.

Before we discuss appealing a denial of a medical certificate, or the special issuance of a medical to a pilot whose physical condition doesn't meet the letter of the rules, there is one age-old piece of advice that every pilot should heed. Don't apply for a medical certificate of a class higher than you need. Unless you're truly aiming at an airline career, why take the risk that the most stringent exam will uncover a problem that would not be discovered if you submitted yourself to the physical for a third-class medical?

I have represented many pilots who, for some reason of ego or ignorance, went in to see an AME for the first time, asking for a first-class medical, only to be bounced by that AME, or by the FAA later. Most of them got through the experience after spending too much money on lawyer bills and extra medical procedures that proved that they were fit to command a Boeing 747, when they would fly nothing larger than a single-engine airplane. If you are a private pilot, or a student who has no plans to fly for a living, stick with the exam for a third-class medical.

Pilots of certain categories of aircraft—namely, gliders and balloons—have had the privilege of self-certifying their medical condition for decades. When you apply for a glider or balloon rating, the application has a little section where you sign your name to a statement that you have no condition which would keep you from flying safely. That's all that there is to it. However, once you fail the physical for a medical certificate, you cannot self-certify until the disqualifying problem is corrected. Gliders and balloons aren't the aircraft of last resort for those unable to obtain a medical certificate after the exam has been failed. Once you are finally denied a medical, you've become an ultralight pilot.

When you undertake a study of the percentage of serious accidents attributed in some manner to pilot physical incapacitation, you come to a startling conclusion. The percentage of accidents in powered aircraft, for which a medical is required, is almost exactly the same as it is in gliders and balloons, and that percentage in both instances is so small as to be statistically insignificant. The possible elimination of the medical requirement for recreational pilots is a small step in the FAA's coming to realize that the pilot's physical condition is almost never a factor in an accident, and when it is, it is always something that a physical exam could not have uncovered in the first place.

Glider and balloon pilots aren't a bunch of suicidal physical wrecks who can't fly airplanes because of their health. The numbers prove the point. Maybe someday, private pilots will be freed from the expense and hassle of medical certificates, but that day won't come very soon—not until there is a statistical group of recreational pilots flying powered aircraft without medicals, and the FAA sees that no increase in the level of risk is associated with the lack of a medical.

Until that day arrives, if it ever does, we'll all need a medical certificate to fly powered aircraft at the level of a private pilot or higher.

CHECK YOURSELF FIRST

Before your your first visit to a physician who has been designated as an AME, get a copy of Part 67 of the FARs. This part sets forth the standards for the three levels of medicals. Read it, and see if there is anything in your physical history or condition that sticks out to you as potentially disqualifying. In this enlightened age, when most sophisticated people have a heightened awareness of their physical fitness, almost anyone can read Part 67 and make this initial determination.

If you have any doubts after reading Part 67, see your own doctor before going to the AME. Explain to her that you want to be a pilot, and take your copy of Part 67 with you. Let your doctor give you her first impressions of whether you have a problem that may prevent you from getting a medical from the FAA. Then, go see the AME.

Explain to the AME, in an initial office conference, without formally applying yet for a medical, what you and your doctor see as a potential problem. If the AME you've chosen won't have this conference, find a different AME. Once you've filled out the application, and submitted yourself to the exam, the AME has no alternative but to defer issuance of a medical to the FAA if you don't meet the published standards. But there is nothing immoral or illegal in sitting down with an AME first, in a conference setting, before officially applying for a medical certificate. In this meeting, the AME can probably determine if you can't pass (few pilots fall into this category), and he can tell you what medical records may have to be reviewed, or what further tests you may have to undergo, in order to get certified.

Then, you can get the records in hand, or take the other tests, and have all the required documentation in order when you go back and formally apply for a medical. That way, if the AME has to defer to the FAA, all your ducks are in a row, and the certification process can be many months shorter, and have a much higher likelihood of eventual success, than if you don't get things together first. At the risk of repetition, if an AME won't go through this step with a potential pilot who is in doubt, or with an aviator who has recently suffered some sort of significant health-related event or illness, that pilot should find an AME who will. Most AMEs are pilots who want to see you certified, but in any profession there is always that odd one who has the "bedside manner" of a caged rattlesnake. Avoid that AME.

A REQUEST BY THE FAA FOR FURTHER RECORDS

Some pilots go to see an AME, pass the exam from that doctor, and get a medical certificate, only to receive a certified letter from the FAA Aeromedical Branch a few weeks later that requests the pilot to produce certain medical records. This letter can't be ignored, or the medical certification that you do have will soon be suspended or revoked. You have to deal with this problem immediately.

At this stage, you need a cooperative AME if you haven't already been dealing with one. You may also need the advice of an aviation lawyer, depending on the severity of the problem that has led the FAA to question your eligibility for medical certification. You should carefully read the FAA's letter, and contact the doctors or hospitals that have the records the FAA wants to see. Start early, since sometimes it may take a few weeks to retrieve the records, especially if they go back more than a year or two.

Once you get the records, don't immediately send them to the FAA. Rather, call your AME, make an appointment, and go to him with the documentation in hand. It is important not to omit this step. You can't hide anything from the FAA, and an ethical AME won't assist in any such scheme. What the AME can do is review the records first. In that review, he'll be looking for errors in the records, inconsistencies which might need further explanation, and real problem areas upon which the FAA might seize to challenge your medical certificate.

If the documentation is not in order, the AME can be of tremendous assistance in putting together a cogent package to send to the FAA, so that you are not caught up in a paper chase caused by incorrect or missing records. Few of us who are not physicians can catch these kinds of problems in medical records. If you send a bunch of paperwork to the FAA that is incomplete, or unintelligible, the review process will just take that much longer. In most of the instances when the FAA requests records from a pilot, the pilot will not suffer any challenge to her medical certificate, and the inquiry will end once the medical data are supplied as requested. But be on the safe side and have the AME review your records first. If there is a real problem uncovered, the AME can start the process, whatever it may be, of performing any additional tests or diagnostic procedures to counter the negative import of what may be in your previous medical history.

WAIVERS AND SPECIAL ISSUANCE CERTIFICATES

Only the Federal Air Surgeon can make an official denial of an application for a medical certificate. If a pilot visits an AME, and does not walk out of the office with a medical in hand, that pilot has not been denied a medical certificate. She can reapply to the same, or a different AME at any time. When the AME feels that a medical cannot be issued, the doctor is generally deferring the decision to FAA aeromedical headquarters.

Then, the FAA folks at Oklahoma City will review the pilot's application, and the results of the physical examination performed by the AME. If this happens, the chances are not good that the pilot will immediately become certified. Because, in most doubtful situations, where the AME thinks that there is a possibility of certifying the applicant, the doctor always has the freedom to call the FAA, explain the question over the phone, and get telephonic permission to issue a medical certificate. This happens more frequently than one would initially imagine.

Some physical or mental conditions which are red flags can be eased over time. One of them is a history of mental illness. Depending upon how the diagnosis and treatment went at the time that the applicant was suffering from a certain malady, the AME may feel comfortable with the person's present state of health, and have no personal hesitancy to issue a medical. With an appropriate phone call, the FAA will most likely approve the issuance, if the AME pleads the case well.

Other disqualifying conditions, such as a history of heart attack, can never be resolved over the phone. In instances of these severe problems, the AME will finish the exam, package the results, and defer to the FAA for issuance or denial of a medical. In the face of a disqualifying condition or history, no AME has the authority to issue a medical without FAA approval. Often that approval is obtainable by phone; sometimes not.

Once the FAA aeromedical people receive the package, they have to decide if the applicant meets the standards prescribed in Part 67. If our pilot meets the requirements, then a medical certificate will be forthcoming in the mail. If not, the FAA goes further into more queries.

If the FAA doctors feel that the particular applicant does not meet the requirements, they have to decide if a special-issuance medical is justified, if a regular medical can be issued with waivers, or if the applicant should be denied a medical. Waivers are routinely issued for certain conditions. To those of us blessed with normal color vision, it is difficult to imagine what it is like for people who are color-blind. Since the recognition of colors plays a part in flying aircraft, from seeing position lights on aircraft at night, taxiway versus runway lights, rotating beacons, and airport markings in the day, Part 67 has always required a pilot to be able to distinguish the colors of red, white, and green. Most AMEs use a color vision test comprised of pages in a notebook like series of pastel-colored displays, in which circles or triangles are apparent to the person with normal color vision, but are invisible to the one whose color vision is impaired.

If you fall into that category, you cannot receive a medical certificate from the AME. Rather, your application will be deferred, and assuming that everything else is satisfactory in your exam, the FAA will set you up for a test whereby you will go to a controlled airport, and be asked to stand in the open, looking at the control tower. The controller in the tower will use the light gun to flash prearranged signals, and you will be asked to tell an FAA inspector, standing there with you, what colors you see. If you pass this test, you will be given a medical with a "demonstrated ability" statement attached, commonly called a waiver. You demonstrated ability to see colors well enough to meet the essence of the Part 67 requirements, even though you failed the color vision test in the AME's office.

For those few who cannot distinguish the light gun colors (generally men suffer color-deficient vision far more often than do women), all is not lost. They can still receive a medical, but it will contain an endorsement that it is not valid for night flight. This means that those pilots cannot fly, as pilots in command, at night, but otherwise may be fully certified as pilots. That certainly isn't the best of all worlds, but it still allows the pilot to fly in daylight, when color recognition isn't vital. That limitation may seem more restrictive than it is. Most private pilots fly very few hours at night anyhow. In over 30 years of flying, I have logged less than 10 percent of my total hours at night. If I had to choose between flying only in daylight and not flying at all, I'd stay out of the dark.

Waivers are also issued quite often for people who have lost the vision of one eye. Since human depth perception normally uses binocular vision for the cues that enable us to tell how far we are from some object, the one-eyed pilot will have to demonstrate her ability to fly an airplane, particularly in landing. An FAA inspector will accompany the applicant on a short flight test. If the pilot can fly without both eyes, again a statement of demonstrated ability will be issued, since the pilot has demonstrated her ability to perform adequately.

One of my friends lost a leg in an accident years ago. His medical certificate was limited to aircraft that did not have only foot-operated brakes, since his artificial leg could not pivot about any substitute ankle joint to apply pressure to the brake pedals. He flew airplanes that have handbrakes, like most of the older Pipers did, and helicopters. Since most helicopters have skid landing gear, they have no brakes at all. A creative solution to the problem.

There are pilots flying who have no hearing at all. Their medicals are noted as not valid for flights into airspace where two-way communication with ATC is required. The same limitations have applied to pilots who cannot speak. Medical certificates have been granted to pilots who are bound to wheelchairs, and who are limited to flying airplanes without foot-operated controls. An example is the old Ercoupe, which does not have rudder pedals (since it was manufactured with an interconnect between the ailerons and the twin rudders) but can be equipped with a hand brake in place of its standard single-foot brake.

If the pilot's physical deficiency is one that cannot qualify for a statement of demonstrated ability, the FAA can give what is called a special-issuance medical certificate. These are granted when the Federal Air Surgeon is convinced that a person, who does not meet the letter of the legal standards, can still be safely allowed to fly. The greatest number of these certificates are sought by people who have a history of heart attack, or open heart surgery. Such a history is automatically disqualifying, per Part 67. However, the FAA has, in the last couple of decades, taken an enlightened view of people with heart disease and heart surgery in their medical histories.

If the FAA can be convinced that the pilot's condition is stable, and that the heart is healthy enough to allow this applicant to be safely in the air, a special-issuance medical is a distinct possibility. Usually the pilot is required to sit out at least a year after the infarction or surgery before the FAA will entertain an application for a special issuance. Then, depending exactly on the nature of the history, the condition, and the treatment that was rendered at the time, an exhaustive regimen of tests will be specified. This regimen will include a stress test, an EKG, and whatever else the aeromedical staff thinks is needed to ascertain this applicant's cardiac health.

If the test results are approved, the FAA will give the pilot a special-issuance medical. It may contain certain limitations, such as a shortening of the period of validity from what would be normal for a pilot who doesn't have to go the route of special issuance. Perhaps a third-class certificate will have its expiration stated at 6 or 12 months from issuance date, rather than the normal 24 months. Perhaps additional stress tests or EKGs will be required at renewal, with the test results being submitted to the FAA as a condition of continued validity of the medical. A special-issuance medical can contain any additional limitations or requirements that the FAA deems necessary to monitor the pilot's state of health.

Most experienced AMEs will quickly tell you that the key to a successful application for a special-issuance certificate is a well-prepared package to accompany the results of the normal physical that the AME will give you at the time you see him. The FAA, like all large organizations, gives a great deal of weight to how well your case is documented. If you've had a history of cardiac disease or treatment, cancer, high or low blood pressure, or any of the other conditions or diseases specified in Part 67 as disqualifying, you've got to have your application well documented. The assistance of a helpful AME is absolutely vital. Should you fall into any of these categories, don't automatically assume that you can't fly.

Persons with a history of hypertension, also known as high blood pressure, are commonly certified. Once these patients take appropriate medication, and cardiac and renal tests disclose no impairment to the heart or kidneys, their chances of getting a medical are quite high. At one time, the FAA had a very limited list of approved mediation for high blood pressure. The list has since been expanded significantly. If you suffer from hypertension, see the AME first, before you present yourself for your physical exam. Ask if the medication you are taking is on the approved list. If not, then ask your regular doctor if you can be switched to medication that the FAA accepts. If so, and if there are no other problems, you'll probably soon be in possession of a medical.

Special-issuance certificates get more difficult to obtain as you apply for higher levels of certification. The FAA tries to balance the freedom of personal aviation with the concurrent need to protect others in the sky and on the ground when it comes to special-issuance medicals. As you can readily conclude, a different tack is taken if the applicant is seeking a second- or first-class medical. With the two higher levels of medical certificates, the FAA must assume that the pilot wants to fly commercially, carrying passengers or property, or both, for hire. In most all areas of the law, paying customers are granted a higher level of protection from those whom they hire to provide transportation. The FAA feels no differently in deciding whether to grant a special-issuance medical.

A rather famous air show pilot had his medical certificate revoked in the early 1990s, for alleged cognitive deficiencies. He appealed the revocation, using the procedure that we'll examine in the next section of this chapter. He lost all the appeals, and was forced to be grounded in the United States. During the time that he had no valid medical in the states, he went to Australia and was given Australian pilot and medical certificates. With these in hand, he could perform his airshow routine anywhere in the world except in his own country.

Intense public and media pressure continued to be applied to the FAA to restore this legend in the airshow world to flying status in the United States. After about 2 years, the FAA relented, and allowed him to take another battery of mental and cognitive tests. At the completion of these exams, he was given a special-issuance second-class medical certificate, with a limitation that he could not carry passengers for hire. Then, he was able to resume his airshow work in the United States, since his act doesn't involve anyone else's being in the airplane with him.

However, his medical is limited, since it is not valid for carrying passengers for hire, as a second-class certificate is otherwise.

APPEAL OF DENIAL OF A MEDICAL CERTIFICATE

If an application for a medical certificate has been formally denied by the Federal Air Surgeon, who is the only person who has the power to make a formal denial, and if the pilot cannot convince the FAA to grant either a waiver or a special-issuance medical, the only avenue left for that pilot is to appeal the denial. Appeals of denials of medical certificates are filed with, and heard by, the National Transportation Safety Board (NTSB), in much the same manner as are regulatory enforcement cases. This area of aviation law is thoroughly discussed in Chapter 6.

The procedure involved in a medical appeal is not much different from that in a regulatory appeal. The NTSB assigns one of its administrative law judges to conduct an oral hearing, in which both the pilot and the FAA present witnesses, expert testimony, and other traditional forms of evidence. Even though the format and legal procedure are much the same, there are two significant differences between appealing a case involving the alleged violation of a regulation and one involving the denial of a medical certificate. Of these two differences, we'll shortly see that one of them generates the other.

When a pilot is appealing a denial of a medical, the pilot's medical condition or illness is the focal point of the entire endeavor. In order to present much of a case to the law judge, it will be critical for the pilot to employ qualified physicians to examine her, and then to testify in the hearing about their findings and about her suitability for medical certification. You can "take it to the bank" that a pilot will not be able to overcome the findings and testimony of the FAA doctors without very competent and compelling testimony from doctors employed by the pilot.

Since this quantum of expert examination and testimony is vital to the pilot's chances of success, the result is that appeals of denials of medical certificates are very expensive for the pilot involved. Unless you are willing to spend lots of money, measured in the thousands of dollars, don't waste smaller amounts even trying to appeal. The costs can mount up fast for extensive medical examinations and tests, and for the legal fees incurred for a lawyer to properly prepare and present the pilot's case. Naturally, there is an exception to every general rule, and perhaps there is a medical case which is so simple that it can be done on a small budget. In my practice, since 1975, I haven't seen that case yet, but it still might be out there.

Also, any pilot considering the choice of whether to appeal a denial of a medical certificate must face up to the sobering fact that very few of these appeals are successful. The NTSB law judges will, most often, defer to the judgment of the FAA medical folks, since it is their job to evaluate pilots' suitability for certification, and they are quite competent at it.

After the oral hearing before the law judge, the loser of that proceeding has the right to appeal further to the entire National Transportation Safety Board. These are usually paper appeals, which means that the parties file briefs with the board, which then considers the limited appellate issues without taking testimony. However, if any oral argument is presented, the pilot must be prepared to send his lawyer to Washington, D.C., where the NTSB sits. A law judge comes to you, and holds the initial hearing at a location reasonably convenient for you and your experts. But when you appeal to the full board, you and your entourage go to Washington. Not too difficult for someone from Richmond, Virginia; it's a wholly different matter for someone from Los Angeles.

Even if the pilot wins the initial hearing before the administrative law judge, the FAA can appeal that decision to the full NTSB. In medical cases, the FAA probably will; so the pilot needs to budget enough money for this additional step. If that much isn't affordable, think long and hard about whether it's worth it to even start the appellate process.

If you lose the appeal to the full board, it is still possible to appeal further into the federal court system. Unless you have a legal defense fund behind you, as did the airshow pilot about whom I spoke a few paragraphs ago, or lots more money to invest in the quest, forget the court appeals. The chances of success are very low to begin with, and they get exponentially lower with each succeeding step of the entire appellate process.

After reading all this dismal material about the costs and low success rates of appeals of medical certificate denials, remember that denials are rare. Most all applicants who should be flying are able to get a certificate, or at the least they either get one with a waiver or go through the route of special issuance. The FAA makes a few mistakes, as it did with the famous airshow pilot; but most of the time when an application for a medical certificate is formally denied, it ought to be. That is particularly so if the applicant has presented a well-documented attempt to get either a waiver or a special-issuance certificate.

Do your best to keep yourself in reasonable physical condition, and communicate with your AME if you have doubts about your eligibility for certification before you actually take the physical exam. Lastly, keep in mind that almost all pilots who really want and should have a medical will get one.

11

Conclusion

IT IS MY FERVENT DESIRE THAT THE FOREGOING MATERIALS ON aviation law have broadened the understanding of those of us who are engaged in the general aviation industry about the legal system and how it affects our businesses. It should be made patently clear again at this point that this book is not intended to provide particular legal advice to any person in any specific situation. It was written with the desire to heighten the level of awareness within the aviation community about the law, but no singular work can ever presume to give legal advice to one person, or group of persons, about a situation with which they are faced, or which they might encounter in the future. The services of competent counsel must be engaged to analyze the tremendous number of questions and issues that arise in every legal query and controversy. Readers are encouraged to form a relationship with an attorney or attorneys who can serve their needs and take every advantage of the opportunity that exists to receive good legal advice from counsel.

Just as physicians are able to practice some of their best medicine in the arena of prevention of disease and debilitation, lawyers can often guide clients through the maze of the legal system and its implications upon those clients' businesses before problems occur. But the physician has a definite advantage. In many situations, physical symptoms appear when a particular malady is still curable. Lawyers do not often enjoy a similar edge. To the uninformed, a legal problem can deteriorate quickly, and have no outwardly visible manifestations of an ill-toward result until it is virtually too late for even the most skillful attorney to turn the tables in a client's favor.

The reader of this book has gained only what a book of this general nature can provide, and that is an awareness that our increasingly complex legal system affects all of our endeavors. Armed with that foreknowledge, the wise will use it, not to form their own legal opinions, but to realize that it is time to see their attorneys and get the advice and guidance that are called for in the situation.

Rather dramatic changes have recently occurred, and will continue to occur, in the regulatory scheme of aviation, both as promulgated and implemented by the Federal Aviation Administration. In 1989 an entirely new level of pilot certificate was created—the recreational pilot. In 1995, a very extensive rewrite of Part 61 of the FAR was proposed, and as of this writing, it is still in the rule-making process.

Perhaps the most sweeping change in aviation law was the 1994 enactment by Congress of the 18-year statute of repose (GARA), giving general aviation manufacturers some badly needed immunity from product-liability suits.

Everything possible must constantly be done to strive to increase the safety of general aviation. It is no surprise to most of us that there is still a large percentage of the general population who consider light airplanes to be dangerous, and the flying of them to be something done by those who do not highly prize their lives and limbs.

Any improvement in the safety of general aviation will go great measures toward making the industry not only safer for those of us who participate in it but, perhaps just as importantly, more acceptable to the rest of our neighbors who are not directly involved but who certainly indirectly benefit from this important element of our national transportation system.

The reader has been acquainted with some ideas concerning the ownership of aircraft and some caveats relating thereto. Every aircraft manufactured must eventually be purchased by someone, or the manufacturer will not remain in business. As the cost of flying constantly increases, more creative approaches will undoubtedly come along to the problem of affording aircraft ownership. Each one of those ideas will have its own subset of legal ramifications that will need to be properly raised and analyzed.

The insurance industry has gone through tremendous upheavals through the 1980s and 1990s. Just a few years ago we were all hearing about a liability crisis of such severe proportions that many national pundits were hypothesizing that the day would soon come when many ordinary individuals, businesses, and perhaps even governmental entities would be unable to procure insurance to cover even the most basic liabilities they face. Thankfully those dire predictions did not come true during the mid-1980s, and as of the mid-1990s the insurance market has softened substantially and, in some particular areas, has even become classified as a buyer's market.

But unquestionably, the insurance industry will see more turnabouts, because it has almost always been a cyclical business at best. In order to be adequately informed and to be able to make reasoned decisions, virtually every participant in aviation must be aware, and keep a finger on the pulse, of the insurance coverages and premiums available. Many companies are writing new forms of policies in what is hopefully more plainly worded language, in an attempt to demystify the process to the insured and other people who come into contact with those documents. Like almost everything else in our fast-paced world, the arena of aviation insurance will probably not be the same 10 years from now, because it is certainly not now the same as it was 10 years ago.

Of the entire body of aviation law as it affects the general aviation pilot, there is probably no more important segment than that of the FAA's enforcement policies and procedures. During the 1980s, the FAA went on what some have described as an enforcement binge, seeking to violate and prosecute certificate holders for the most sublime and minor technical breaches of the Federal Aviation Regulations. Much was said in Chapter 6 about that subject, and it will certainly not be repeated in these closing remarks.

All aviation certificate holders need a working knowledge of the FAA enforcement policies and practices and, most importantly, of their rights when confronted with that system, because as it now stands a violation of a FAR might become as common as the traffic violation. However, we all know that the consequences are far more severe for a pilot who suffers a certificate suspension than for the automobile driver who is fined for a speeding infraction. There are certainly serious automobile offenses, such as driving under the influence of alcohol or some other drug, which rightfully carry significant penalties; but as things go, a pilot has far more to lose, and will suffer a far greater penalty in proportion to the seriousness of his conduct, than the individual who casually violates one of the more minor traffic laws.

The effects of a certificate action will follow a pilot for years and, if he is a professional aviator, perhaps for the rest of his career. This entire area has become so volatile and so emotionally heated over the last few years that all who operate within the aviation industry need to be aware of, and keep themselves up to date on, what the FAA is doing relative to the enforcement of its regulations.

We examined the principles of negligence liability as they apply to most businesses and individuals involved in the industry, other than manufacturers, who deserved and received separate treatment.

This is a litigious society. People file lawsuits today for some of the dumbest and slightest of reasons. But even outside of the ridiculous cases, all of us need to have heightened knowledge of the reality that we are liable for our actions.

Virtually every conceivable business and professional endeavor is aware of and concerned about the exposure that it faces to its customers and clients in the area of negligence liability. This area of the law is so broad that what was covered in this book probably does not even rise to the level of having scratched the surface. But we must begin somewhere, and I have strived to acquaint you with the salient subjects of which you need to be aware, so that competent legal advice can be sought concerning specific situations.

Lastly, I have covered the area of product liability, and the volume of the discussion bears a good relationship to the importance of it. Probably no other area is affecting general aviation today as much as the problems related to product liability. Just a few years ago, this entire subject was something that was discussed in law schools, in scholastic legal works, and in a few court opinions. Today, "product liability" has become a household term, although probably not well understood by a good number of the people who mouth it.

But even without having an in-depth understanding of the law of product liability, virtually every educated and sophisticated person is aware that this area of the law is affecting virtually everything she purchases, from a toaster to an airplane. Because the theories involved in product-liability cases are relative newcomers to the American legal system, this area of the law is one of the most rapidly changing and evolving. An attempt was made to prognosticate the future and, as with all such activities, a look back in a few years might prove that the law has taken a different direction than anyone could have forecast at this juncture in time. At least our industry had accomplished something that is very special, since we now have a federal law which is aimed directly at protecting general aviation manufacturers, and helping revitalize the industry.

In conclusion, the aims of this work will be served if the reader gains some useful information and enjoys the increase in knowledge to even a small fraction of the pleasure I had in preparing, writing, and communicating it. In the last analysis, the results of cases tried in our legal system will reflect the safety level of the aviation industry. To that end, the reader is wished happy landings, fair weather, an ever-present tail wind, and a safe and enjoyable conclusion to every flight.

Index

A

A&P mechanic, 27–28
accidents (see also negligence liability), 3, 4, 174
administrative law, 10
administrative action, FAA enforcement, 123–126
adversary legal system, 4
affirmative defenses, negligence liability, 139
Aircraft Owners and Pilots Association (AOPA), 37
aircraft ownership, 31–57
Aircraft Registry, FAA, 35, 36, 41, 42, 43
aircraft title insurance, 37–38
airframe mechanic, 27
airline transport pilot certificate, 19–21
airplane, 16
airship, 16
airworthiness directives, product liability and, 195
amateur built aircraft (see homebuilt aircraft)
American legal system, 1–6
  adversarial nature of, 4
  defendant in, 6
  English common law background of, 2
  judge and jury in, 3–4
  justice, fact and fiction, 4–6
  plaintiff in, 6
  precedent in formulation of, 1
American legal system (*Cont.*):
  State vs. Federal law, 2–3
appeal brief, NTSB appeals, 119
appeals to NTSB, 115–120
appellate body (see NTSB)
Appendix A (owner performed maintenance), 54–57
approved repair stations, negligence liability and, 159
assumption of risk, negligence liability, 140
ATC, 15
attorneys, 4–6
attractive nuisance doctrine, 171
aviation law, State vs. Federal, 2–3
Aviation medical examiner, 12
Aviation Safety Reporting System (ASRS), 108–115
  condition of entitlement, 112
avionics, 48

B

balloons, 16
bill of sale, 38–40
breach of duty, 127, 130–133
buy and sell terms of co-ownership agreement, 49
buying an aircraft, 31–42
  bill of sale, 38–40
buying an aircraft (*Cont.*):
  contract of sale, 34–35
  prepurchase inspection, 32–34
  promissory notes, 41
  sale-leaseback arrangements, 166–169
  security agreements, 41–42
  title search, 35–37

C

categories of aircraft, 16
certificate action, 98
  notice of proposed, 98–105
  notice of, 105–107
certificate of registration, 43–44
certification, 8, 11–24, 65
  currency requirements, 24–26
  flight reviews, 24–26
  revocation of, 98–105, 107–108
  suspension of, 98–105
certificated flight instructor, 21–24
Civil Aeronautics Administration (CAA), 8
civil penalties, 95
  FAA enforcement and, 121–123
  order of, 122
closing arguments, NTSB appeals, 118
clubs, co-ownership, 43–54

co-ownership of aircraft, 43–54
buy and sell terms, 49
doctrine of joint or common enterprise, 175
flying clubs, 43–54
informal, among individuals, 47–51
insurance, 45
loans and financing, 50
negligence liability and, 174–176
written agreements for, 47–51
Code of Federal Regulations, Part 821, NTSB role, 115
combined single limit liability insurance, 86–87
commercial coverage insurance, 79
commercial pilot certificate, 17–19
common enterprise, 175
common law, English, 2
comparative negligence, negligence liability, 140–142
compensation, 3
complaints, 116
completed operations coverage, 80
condition of entitlement, ASRS, 112
conflicts of laws, 3
Congress, 9
consumer expectation test, product liability and, 188–189
contract of sale, buying aircraft and, 34–35
contributory negligence, negligence liability and, 140–142
corporate flying clubs, 52–54
correction, letter of, FAA enforcement, 124
cosmetics, aircraft, 47
courts, 9
crashworthiness, theory of, product liability and, 202–206
currency requirements, 24–26

D

damages, negligence liability, 134–138
general damages, 135–136
loss of consortium, 136
preponderance of evidence in, 138–139
reasonable doubt and, 139
special damages, 134–135
defendant, 6
Department of Transportation, 8
doctrine of joint enterprise, 175
duty to defend clause, insurance, 89–90
duty, negligence liability, 127–130
breach of, 130–133

E

emergency orders, NTSB, 107–108
enforcement, FAA, 8, 10–11, 95–126
administrative action, 123–126
Aviation Safety Reporting System, 108–115
burden of proof rests with, 116
certificate action, 98
civil penalties, 95, 121–123
complaints to NTSB from, 116
flight standards district office (FSDO), 95
letter of correction, 124
letter of investigation, 96–98
Miranda decision and, 96
National Transportation Safety Board appeals, 115–120
notice of certificate action, 105–107
notice of proposed certificate action, 98–105
order of civil penalty, 122
order of suspension, 106
prima facie case, 118
enforcement, FAA (*Cont.*):
rules of evidence in appeals to NTSB, 117
stale complaint rule, 104–105
U.S. Circuit Court of Appeals, 120
U.S. Supreme Court, 120
warning notice, 124
English common law, 2
entitlement, condition of, ASRS, 112
entrustment, negligent, negligence liability and, 149
evidence, 4
NTSB appeals, 117, 118
preponderance of, negligence cases, 138–139
executive branch, 9
expressed warranty, product liability and, 181–182

F

failure to warn, product liability and, 193–195
Federal Aviation Act of 1958, 3, 10
Federal Aviation Administration, 7–29
airman certificates and ratings, 11–24
currency requirements and flight reviews, 24–26
enforcement, 8, 10–11, 95, 126
Final Notice of Rule Making, 9
functions of, 8–11
history of, 7–8
maintenance certificates, 26–29
National Transportation Safety Board and, 11
Notice of Proposed Rule Making, 8–9
policy making by, 10
rule making, 8–9
Federal aviation law, 2–3
State laws in conflict with, 2–3

Federal Aviation Regulations (FARs), 9, 10
Final Notice of Rule Making, 9
financing, 41–42
first-class medical certificate, 19
fitness, implied warranty of, 184
fixed base operator:
attractive nuisance doctrine, 171
business invitees, negligence liability to, 172
flight instructors and negligence liability, 169–170
flying and nonflying (ground) areas, 172
insurance, 79–80, 87–89
licensee, negligence liability and, 171
negligence liability and, 127, 145–150, 164–173
sale-leaseback arrangements, negligence liability and, 166–169
trespassers, negligence liability and, 171
vicarious liability of, 169
flight instructors:
certified, 21–24
negligence liability and, 151–157, 170
flight reviews, 24–26, 81
flight schools, negligence liability and, 152–153
flight standards district office (FSDO), 95, 96
flying area, FBO, 172
flying clubs, 51–54
doctrine of joint or common enterprise, 175
negligence liability and, 174–176
unincorporated vs. incorporated, negligence liability and, 51–54, 175–176
foreign jurisdiction, State aviation laws and, 3
Freedom of Information Act (FOIA), 103
fuel suppliers, negligence liability of, 157–159

G

general damages, negligence liability, 135–136
gliders, 16
ground area, FBO, 172
gyroplanes, 16

H

hangar keepers' liability coverage, 79–80
helicopters, 16
homebuilt aircraft, 59–76
Advisory Circular no. 20–27D, 62
annual inspection of, 71
certification steps, 65–66
airworthiness certificate, 65
registration number, 65, 67
designated airworthiness representative, 61
Experimental Aircraft Association (EAA), 59, 61, 62
fifty-one percent rule, 62–64
commercial assistance, 64
list of approved kits, 63
flight standards district office (FSDO), 61
insuring, 72
operating limitations, 65, 67
purchasing a homebuilt, 74–76
quick build kit, 60
repairman certificate, 66
restoring older airplanes, 72–74
selling homebuilt aircraft, 74–76
contract of sale, 75–76
protecting the builder, 75
strictly liability in, 75
homebuilt aircraft (*Cont.*):
technical counselors, EAA, 62
ultralight vehicle, 71

I

illegal use of aircraft, insurance and, 91–93
immunities, 103
implied warranties, product liability and, 184
implied warranty of fitness, 184
implied warranty of merchantability, 184
industrial aid insurance use restrictions, 79
informal conference, 102
initial decision, NTSB appeals, 118
inspection authorization, 28–29
negligence liability and, 162–163
inspectors, 10
instrument flight rules (IFR), 16, 20
instrument rating, 16
insurance, 77–93
co-ownership and, 45
commercial coverage, 79–80
completed operations coverage, 80
duty to defend clause, 89–90
fixed-base operators, 80
hangar keepers' liability coverage, 79–80
hull coverage, 83–84
illegal use of aircraft, 91–93
industrial aid use, 79–80
liability coverage, 84–89
pilot warranties, 80–82
pleasure and business use, 78
premises liability coverage, 79
product liability coverage, 80
renter pilot coverage, 87–89
stated value coverage, hull, 83

insurance (*Cont.*):
subrogation, 90–91
total loss, 83
use restrictions, 77–80

J
judges, 2, 4
judicial process, product liability, 199
jury, 3–4
justice, 4–6

K
keepers of premises, negligence liability and, 170

L
last clear chance doctrine, negligence liability and, 142–143
lawyers (see attorneys)
leaseback agreements, 166–169
legislative branch, 9
lessors of rental aircraft, negligence liability and, 145–151
letter of correction, FAA enforcement, 124
letter of investigation, FAA, 96–98
Miranda warnings and, 96
response to, 96–98
liability (see also negligence liability; product liability), 3
vicarious, 158, 169
liability insurance, 84–89
combined single limits, 86–87
per person limitations, 85–86
licensee, negligence liability and, 171
licenses (see certification; ratings)
liens, 36
lighter-than-air, 16
limited warranties, product liability and, 183
loans, 41–42
loss of consortium damages, negligence liability, 136–137

M
maintenance, 46
homebuilt aircraft, 66–67
owner-performed, 54–57
maintenance certificates, 26–29
negligence liability and, 159–164
master-servant rule, 196
medical certificates, 12, 19, 81
appeal of denial, 213–215
merchantability, implied warranty of, 184
Miranda decision, 96
Monroney, Mike, 35
motions, NTSB appeal, 118
multiengine helicopters, 16
multiengine land airplanes, 16
multiengine seaplanes, 16

N
National Aeronautics and Space Administration (NASA), 103, 108
National Transportation Safety Board (NTSB), 11, 101, 106, 108
appeal briefs for, 119
appeals to, 115–120
appellate function of, 115–120
burden of proof in appeals, 116
closing arguments before, 118
complaints from FAA, 116
emergency orders, 107–108, 115
evidence, 117
Federal Aviation Administration and, 11
Federal codes pertaining to, 115
initial decisions from, 118
National Transportation Safety Board (NTSB) (*Cont.*):
investigatory functions of, 115
motions, 118
notice of appeal, 115
opening statement before, 117
prima facie case in, 118
rebuttal argument before, 118
reply brief, 120
rules of evidence for, 117
negligence action, 128
negligence liability, 127–176
affirmative defense, 139
approved repair stations and, 159
assumption of risk, 140
attractive nuisance doctrine, 171
breach of duty, 127, 130–133
business invitees, negligence liability and, 172
co-ownership and, 174–176
comparative negligence, 140–142
contributory negligence, 140–142
damages, 134–138
defenses to, 139–143
duty to others and, 128–130
fixed-base operators and, 158, 164–172
flight instructors and, 151–157
flight schools and, 152
flying clubs, 174–176
fuel suppliers, 157–159
general damages, 135–136
inspection authorization, 162–163
keepers of premises law and, 170
last clear chance doctrine, 142–143
licensees and, 171
maintenance operators, 159–164

negligence liability (*Cont.*):
master-servant rule, 196
negligence action, 128
negligent entrustment, 149
pilots, 173–174
preponderance of evidence and, 138–139
proximate cause, 133–134
reasonable care-reasonable man standard, 129
reasonable doubt and, 139
renters of aircraft and, 145–151
sale-leaseback agreements, 166–169
self-defense, 139
trespassers, 171
vicarious liability and, 169
negligent design and manufacture, 179–181
negligent entrustment, negligence liability and, 149
nonflying area, FBO, 172
nonowner insurance, 87–89
notice of appeal, 115
notice of certificate action, 105–107
notice of proposed certificate action, 98–99
Freedom of Information Act and, 103
immunities, 103
informal conference, 102
options in response to, 100–104
respondent's role in, 103
Notice of Proposed Rule Making, 9–10, 13

O

opening statement, NTSB appeal, 117
overhauls, 46
owner-performed maintenance, 54–57
ownership of aircraft:
acquisition of aircraft, 31–43
co-ownership, 43–54
ownership of aircraft (*Cont*):
owner-performed maintenance, 54–57
sale-leaseback agreements, 166–169

P

partnerships, 47–50
per person insurance limitations, 85–86
pilot warranties, 80–82
pilots, negligence liability and, 173–174
plaintiff, 6
pleasure and business insurance use restrictions, 78
policy making, Federal Aviation Administration, 10
powerplant mechanic, 27–28
prepurchase inspection, 32–34
premises liability coverage, 79
preponderance of evidence, negligence liability and, 138–139
preventative maintenance, 54–57
prima facie case, 118
private pilot certificate, 15–17
product liability, 177–206
airworthiness directives and, 195
consumer expectation test, 188–189
crashworthiness, 202–206
definition of, 178
expressed warranties, 181
failure to warn, 193–195
General Aviation Revitalization Act of 1994, 177, 195–198
implied warranties, 184
implied warranty of fitness, 184
implied warranty of merchantability, 184
judicial process in, 199
limited warranties, 183
negligent design and manufacture, 179–181
product liability (*Cont.*):
question of fact, 183
risk-benefit comparison, 189–193
service bulletins, 194
state-of-the-art question, 191
strict liability, 186–188
suppliers' liability, 198–202
warranties and representations, 181–186
product liability insurance coverage, 80
promissory notes, 41–42
proximate cause, negligence liability, 133–134

Q

question of fact, product liability and, 183

R

ratings, 16
reasonable doubt, negligence liability and, 139
reasonable man standard, 129
rebuttal argument, NTSB appeals, 118
recreational pilot certificate, 13–15
registration, certificate of, 43–44
rental aircraft, negligence liability and, 145–151
renter pilot insurance coverage, 87–89
repair station certificate, 27
repairman's certificate, 66
reply brief, NTSB appeals, 120
representations, product liability and, 181–186
respondent, 103
risk, assumption of, negligence liability and, 140
risk-benefit comparison, product liability and, 189–193
rotorcraft, 16
rule making, Federal Aviation Administration, 8–10

S
sale-leaseback arrangements, 166–169
seaplanes, 16
security agreements, 41–42
self-defense, negligence liability, 139
service bulletins, product liability and, 194
single engine land airplanes, 16
single engine seaplanes, 16
special damages, negligence liability, 134–135
stale complaint rule, 104–105
stare decisis, 1
State aviation laws, 2–3
state-of-the-art question, product liability and, 191
stated value hull insurance coverage, 83
storage, 45
strict liability, products and, 186–188
student pilot certificates, 12–13
subrogation, 90–91
suppliers' liability for products, 198–202

T
third-class medical certificates, 12, 19
time between overhauls, (TBO), 33
Title 14, Code of Federal Regulations, 9
title insurance, 37–38
title searches, 35–37
  former ownership, 36
  liens, 36
  unreleased security agreements, 36
total loss, 83
trespassers, negligence liability, 171
trials, 3–4
type ratings, 16

U
U.S. Circuit Court of Appeals, 120
U.S. Supreme Court, appeals to, 120
unincorporated flying clubs, 51–52
unreleased security agreements, 36
use restrictions, insurance, 77–80

V
vicarious liability, 158, 169
visual flight rules, 14

W
warning notice, FAA enforcement, 124
warranties, product liability and, 181–186
  expressed warranties, 181
  implied warranties, 184
  implied warranty of fitness, 184
  implied warranty of merchantability, 184
  limited warranties, 183
  question of fact, 183
warranty, pilot, 80–82

About the Author

THE AUTHOR OF *GENERAL AVIATION LAW*, JERRY A. EICHENBERGER, HAS been involved in the general aviation industry all of his adult life. He learned to fly during the summer immediately after his graduation from high school in 1965, and quickly became a commercial pilot and flight instructor, working his way through his undergraduate studies at The Ohio State University as a flight instructor on the staff of the O.S.U. Department of Aviation.

He has held a flight instructor certificate since 1967 and currently is a commercial pilot, rated for both multi- and single-engine land airplanes, helicopters, gliders, and possesses an airplane instrument rating. In the years that he has flown, he has logged approximately 5,000 hours total flying time in everything from a J-3 Cub to transport category aircraft.

Mr. Eichenberger has been engaged in the full-time practice of law since 1975; and is admitted to practice before all of the state and federal courts sitting in Ohio; and, as well, is admitted to the bars of the United States Court of Appeals for the 6th Circuit, and the Supreme Court of the United States of America. He concentrates his legal practice in the area of aviation law, and has represented pilots, airlines, aircraft dealers, manufacturers, maintenance facilities, and fixed-base operators in litigation, as well as serving as counsel to many of those businesses outside of the courtroom.

Mr. Eichenberger is a partner at the Columbus, Ohio, law firm of Crabbe, Brown, Jones, Potts & Schmidt, where he co-chairs the transportation law practice section.

He has served as an adjunct professor at The Ohio State University, where he has taught a course in aviation law within the Department of Aviation. He is a resident of central Ohio, where he lives with his wife of over 25 years (who is also a certificated private pilot), and his daughter.